Virtual Reality Excursions
Study Guide

Michael Kelly, Michael Ort, Steve Semken,
Jay Shiro Tashiro
Northern Arizona University and Dine College

Executive Editor: Dan Kaveney
Special Projects Manager: Barbara A. Murray
Development Editors: Daniel Schiller; Laura Pople
Media Project Editor: Charlene M. Barr
Assistant Managing Editor, Media: Amy Reed
Supplement Cover Designer: PM Workshop, Inc.
Art Director: Joseph Sengotta
Manufacturing Buyer: Diane Hynes

© 2000 by Prentice-Hall, Inc.
Upper Saddle River, NJ 07458

All rights reserved. No part of this book may be
reproduced, in any form or by any means,
without permission in writing from the publisher.

Printed in the United States of America

10 9 8 7 6 5 4 3 2 1

ISBN 0-13-096262-7

Prentice-Hall International (UK) Limited, *London*
Prentice-Hall of Australia Pty. Limited, *Sydney*
Prentice-Hall Canada Inc., *Toronto*
Prentice-Hall Hispanoamericana, S.A., *Mexico*
Prentice-Hall of India Private Limited, *New Delhi*
Prentice-Hall of Japan, Inc., *Tokyo*
Prentice Hall of Singapore Pte. Ltd., *Singapore*
Editora Prentice-Hall do Brasil, Ltda., *Rio de Janeiro*

Preface

Welcome to Virtual Reality Excursions, a set of three virtual worlds in which you can study complex environmental problems. This manual provides a quick review of why Virtual Reality Excursions (VRX) was developed and how you can use it to conduct research in simulated environments. The virtual worlds include: (1) a landfill, (2) Yucca Mountain, which is being considered as a national repository for nuclear wastes, and (3) a virtual worldsthat includes a coal formation simulation and modern settings such as a coal power plant management model. Each of these three virtual environments has four research simulations, and each simulation has a research problem that can be used as a laboratory exercise, homework assignment, or demonstration during class.

As you work on the research problems, you will enhance your knowledge about environmental problems and how to study these problems. Importantly, VRX was designed to help you develop abilities to conceptualize interactions within complex environments, design and carry out research to study these interactions, compare your results to those of classmates and scientists, and synthesize a deeper understanding of the many processes shaping the environments of earth. In short, you will be doing science to learn science.

The manual has five chapters.

- **To the Student** ....

    The first chapter is a study guide that helps students use the research simulations effectively to learn science and think more deeply about what scientists actually do when they study complex systems.

- **Getting Started** ....

    Chapter 2 provides an overview of the system requirements for VRX, with instructions for installation and a description of the virtual worlds.

- **Land Full of Landfills**..

    This chapter contains the landfill research simulations. The virtual landfill has a number of complexities that are similar to many existing landfills. The site has monitoring wells and equipment you can use to sample these wells. You will be able to conduct research on groundwater flowing near and through a landfill, examine possible contamination of groundwater by landfill leachate, and study movement of gases through soils near the landfill.

- **Yucca Mountain – To Store or Not to Store Nuclear Wastes**...

  The Yucca Mountain research simulations in Chapter 4 provide an opportunity to conduct research related to the transportation and storage of nuclear waste generated by nuclear power plants. Yucca Mountain is a real site and has been proposed as the national repository for nuclear waste. You will be able to study the geology of the area, groundwater movements near the site, potential volcanism in the Yucca Mountain area, and the problems associated with transportation of nuclear waste from sites all over the United States.

- **Mire to Fire – Coal Power** ....

  In this virtual world of Chapter 5, we have created a Late Cretaceous mire in which you can study the initial stages of coal formation. Then travelling forward in time some 65 million years, you can study coal beds formed from the mire and decide where to site and how to operate a coal-fired power plant that uses the coal. Research opportunities include collecting and analyzing soil cores from the Cretaceous mire, examining present-day geology and land use in the area where the mire was located, deciding where to site a coal-fired power plant, and conducting optimization studies of running the power plant.

If you're the kind of person who really likes to jump into software and explore a bit, skip to Chapter 2, which gives you the basics for **Getting Started**. On the other hand, if you want to get a sense of what **VRX** will do and why it was designed as a set of research simulations, Chapter 1 is addressed **To the Student** and provides an overview of how you can use **VRX** to improve your learning. If you like compromise, throw the CD-ROM into your computer while you start reading this manual. The instructions on the CD-ROM packaging are a condensed version of Chapter 2's **Getting Started**, so you can quickly start installation.

Table of Contents

Chapter 1: To the Student..1

Chapter 2: Getting Started..27

Chapter 3: Land Full of Landfills...33

Chapter 4: Yucca Mountain — To Store or Not to Store Nuclear Wastes..85

Chapter 5: Mire to Fire — Coal Power..139

To The Student

INTRODUCTION TO VRX

VRX is a set of **mind tools** that can help you learn about complex interactions in the environment by doing research. In this chapter, we provide some guidance about how to use the research problems to learn science and think more deeply about what scientists actually do when they study complex systems. There are four sections following this introduction.

> Doing Science, Learning Science!............................. 3
> What's the Question?..14
> Who's Responsible for Teaching and Learning?............17
> Choose a Pathway to Learning!..............................21

First, we show how you can **learn** science by **doing** science. There are three virtual worlds for you to explore. Each allows you to be a scientist and to conduct research. As you conduct research, you learn. That's Doing Science, Learning Science. One of the virtual worlds is a landfill, and another is a depository for nuclear wastes. As students, you almost never get a chance to study these kinds of environments and their associated problems. Even if they are in your neighborhood, access is often restricted or dangerous, and even if you can get into such a site, the research possibilities are extremely limited.

The third virtual world is a coal-fired power plant. Again, students can seldom study these systems up close and personal. More importantly, in this virtual world **VRX** allows you to study the formation of coal as a fossil fuel, and then study the siting of a power plant that takes advantage of coal deposits. You basically get to travel through geological time, not easy to do in most undergraduate settings. The virtual environments bring real-world problems within your reach. These virtual worlds are easy to get into, easy to explore, and provide realistic research opportunities.

After talking about learning science by doing science, we outline how each research simulations are organized with virtual worlds. For the most part, the best scientists ask questions about how things work, regardless of whether the things are chemical reactions, physical processes, organisms, ecosystems, or systems that comprises an environment. Usually, the questions scientists ask are difficult to answer, but a key point is that science begins with someone asking, "What's the Question?" By posing questions and then trying to answer them by following the scientific method, scientists create knowledge. We will show you how the research simulations provide some insight into what scientists actually do, and so help you do science to learn science.

In this chapter, we also challenge you to take charge of your own learning. We ask you to think about who is the teacher and who is the learner. **VRX** is a new kind of learning tool, with a virtual world that contains research opportunities, but also has electronic texts that you can use for reference. Like any set of tools, their effectiveness depends on how they are used. You have to decide *Who's Responsible for Teaching and Learning*.

Finally, this introductory chapter looks at how people learn. In truth, not everyone learns in the same way. If you are going to take charge of your learning, it is a good idea to know the most effective ways you learn. We offer some suggestions for finding learning strategies that are likely to work for you. There are many different pathways to be successful in learning a subject area. We are pretty sure that doing science is a good way to learn science, but there are still different approaches you can take within the virtual worlds. In the end, successful learning requires that you *Choose a Pathway to Learning*.

DOING SCIENCE, LEARNING SCIENCE!
- You're the Scientist
- So, What is Science?
- Research Worlds and Resource Materials

You're the Scientist

In the virtual environments of **VRX**, you are the scientist. When you enter the software, you are in the Virtual Office, which is your workspace and also the gateway to the virtual environments and their research simulations. The Virtual Office is shown in **Figure 1.1**. Note the Laptop Computer on the desk in the middle of the Office. This laptop provides you a way to log in and choose the environment and simulation that you want to explore. Clicking on the rotating question mark under the desk will display more information about the Office settings.

Figure 1.1

Also note the Bookshelf to the left of the desk and the Simulation Selector to the right of the desk. The Bookshelf provides an immediately accessible source of reference books that can be selected and read on your computer screen. The Simulation Selector allows you to click and drag a chip to the reader at the top of the selector, which opens the virtual environment and the research simulation you chose when you logged in on the laptop.

You can set yourself up in the office, and begin work as a scientist, reading the electronic books or travelling and working in the virtual environments. As a scientist you have to assume the responsibilities of scientists, and so try to do what scientists do when they study the complexities of the world. We can start with a fairly simple definition and say that "*science is the body of knowledge generated by people who apply the scientific method to studying the world.*" Scientists are those people who use the scientific method. As you will later see, we may want to expand on the definition of science and what scientists do, but the important point now is that you will be using the scientific method in **VRX** research simulations.

A major responsibility for you as a student scientist is to be prepared for your research efforts. While scientists are a curious bunch, their curiosity is usually focused and sharpened by studying systems that intrigue them. As you start to do science, you will need to focus on systems and learn something about them. Think about this for a minute. You start looking at the environment, and you note right away that even the simplest of earth's environments are pretty complex. You certainly could start asking questions and try to answer them. However, people have been looking at their environments for a long time, so why not examine what others have found. Now, the truth is that there are few truths, and new discoveries in science lead to changes in how we thought environments were structured. So, there are lots of opportunities for you to contribute to science and change what is recorded as knowledge in research papers and textbooks. The important issue is for you to think about what you have to know in order to ask meaningful questions. Ask yourself what you already know, then ask what you want to know, and, finally, figure out what questions about your area of interest have still not been answered.

Of course, you, as a student scientist, have to start somewhere. For **VRX**, you have to know something about the virtual environments in which you will be conducting research. You will see in the laboratory materials presented in Chapters 3-5 that we give some clues about what prior knowledge you need to begin a research simulation. Most of the information you need will be available in the electronic books found in the Bookshelf that we have placed in the Virtual Office. However, you should also connect the research you are conducting to the courses you are taking. If **VRX** is assigned in a particular course, look at your syllabus and figure out why you use **VRX** in certain parts of the course. What are you studying? What portions of your textbook are related to the research you are doing in a particular virtual world? If you are just exploring the simulations on your own, you might find it useful to get an introductory environmental sciences textbook as a companion to the electronic books in the *Bookshelf*.

So, What is Science?

Well, you are the scientist and you have the responsibility to know something about what you are going to study. We can now turn to discussing how **VRX** provides a better opportunity to do more complex science than the kinds of activities you find in most undergraduate laboratories.

You have probably read quite a few science textbooks and found the scientific method presented in a number of ways, but let's start with a fundamental question, "What is science?" We often use a definition of science provided by Goldstein and Goldstein (1978) in an interesting book entitled, "How We Know." To simplify somewhat, these authors say that science is an activity that has three characteristics (see p. 6 in Goldstein and Goldstein, 1978):

- Science is a search for understanding.
- Such understanding is constructed from general laws or principles.
- We can test the laws and principles by experimentation.

In a simple sense, we observe the real world and try to make sense of the information we have. We feel that part of the beauty and elegance of science is that, as a way of knowing, it is a never-ending story of discovery. This process of discovery is governed by the three characteristics listed above and operationalized in what we call the *scientific method*. However, as a philosopher of science friend of ours once pointed out, "Science is what scientists do." In other words, she meant that there is no single *scientific method* that every scientist follows. Scientists construct a pathway of inquiry within a process broadly described as the scientific method.

Usually, science texts treat the scientific method as a series of steps. **Figures 1.2** and **1.3** show fairly typical summaries of how the scientific method is presented in science textbooks. **Figure 1.2** presents a more elementary version, while **Figure 1.3** shows a presentation of the scientific method for an undergraduate audience that has some training in statistics. However, **Figure 1.4** is a better description of what scientists actually do. Take a look at these three figures and compare the information they contain. Then, focus on **Figure 1.4**. In this figure, we show what we call conceptual looping, which was first described by one of the authors (Tashiro) and a colleague from Tufts University, Dr. Jan Pechenik. **Figure 1.4** is modified from a paper written by Pechenik and Tashiro.

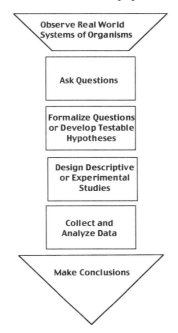

Figure 1.2

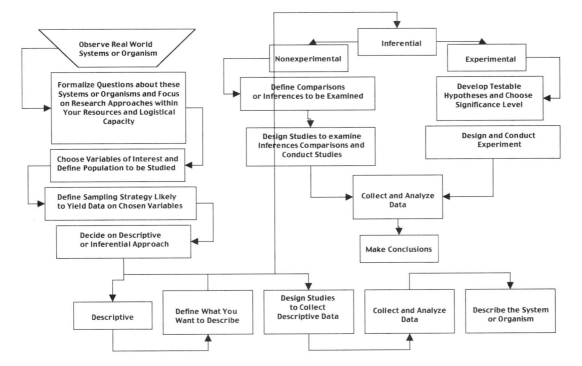

Figure 1.3

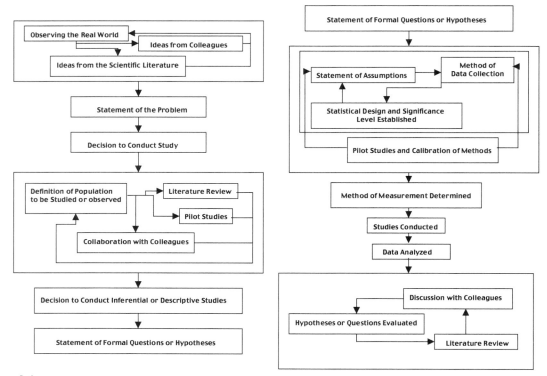

Figure 1.4

Go out, and talk to scientists. Follow them around, and watch what they do. You will seldom find their work as dull and linear as the series of steps shown in **Figures 1.2** and **1.3**. Scientists tend to do what we call conceptual looping. That is, they loop through a series of observations, reading the literature on the system they are studying, talking to colleagues, and working on pilot studies. Only then can they formalize a model of the system and ask specific, meaningful questions. The questions may lead to hypotheses, which are testable statements about the attributes of the system.

Thus, the looping continues. Once a hypothesis has been constructed, scientists begin to conceptually loop through possible designs for studying the system in a way that can falsify the hypothesis. It is important to note here that the best scientific work creates research designs that try to discredit the hypothesis rather than verify it. Support for a hypothesis comes from an inability to falsify the hypothesis.

In any case, the research design is an important element in the process of doing research. The design is shaped by the questions you are asking and the likelihood of getting data you believe can be used to answer the questions. If the questions have been formalized into hypotheses, then the design and the data collected by implementing the design must provide a test of the hypothesis. The research design is almost always coupled to a statistical framework that allows you to make reasonable conclusions about the data you gather. This is another set of conceptual loops. Once scientists choose a research design, they begin thinking about the most appropriate way to analyze the data. The kinds of questions they are asking and the types of data that they plan to collect will shape their decision about what statistical framework to use for analyzing and interpreting the data. **Figure 1.4** provides a representation of the scientific method as a series of interconnected conceptual loops.

Another Important point is that most experienced researchers recognize that most research falls along a continuum from purely descriptive to hypothesis-driven inferential studies. You can see the pathways leading to these two types of studies in **Figure 1.3**. For discussion here, we can simplify a bit and define a hypothesis as a tentative statement about the real world that can be tested empirically. Descriptive studies are not based on hypotheses, while inferential studies are hypothesis-driven.

In environmental sciences, an example of a purely descriptive study is a baseline analysis of an environmental system or set of systems, such an ecosystem, a geological formation, or human constructed system like a landfill. A simple example of a descriptive study would be a research project based only on a broad question, such as, "What are the basic or baseline characteristics of this system?" Take a stream, then you might conduct a descriptive study to analyze the water chemistry, habitat structure, and fauna within this stream system. There are times when we know very little about a system and have to conduct preliminary studies to develop at least some foundation of knowledge about the system. In a sense, we may not know enough to ask meaningful questions or at least not enough to formulate a testable hypothesis about the system. These kinds of studies were much more common in the late 1800s and early 1900s when people were cataloguing plant and animal species, collecting data on freshwater and ocean systems, delineating soil types and geological formations, and literally describing biological, physical, and chemical features of the earth's environments.

In fact, we now seldom conduct purely descriptive studies. Usually research efforts are developed as part of an analysis of some problem. For example, instead of the baseline study, we might be collecting data on an area as part of an environmental impact assessment. Suppose we are asked to complete an environmental impact assessment on an area that has been selected as a site for a residential housing development. For purposes of this discussion, let's say that the site was in upstate New York, in a hilly area on the east side of the Hudson River, about 150 miles north of New York City. The area is rural, with rolling hills, and a pristine stream with trout running through the proposed development site. Prior land use was mostly farming, with no industry, and low population density.

Our goal would be to study the area in detail. There may not be any previous data on some of the habitats, soils, geological formations, surface and groundwaters, or populations of flora and fauna. We are collecting new and diverse data so that we can describe the area and decide if the housing development is likely to have adverse impacts on the area. We are likely to have a number of questions that help us decide what to study, such as: (1) What are the sensitive ecological habitats in the area? (2) What is the water quality of the stream running through the site and what are the stream habitats and their community structure? (3) What are the soils at the site and how will disruption during building shape soil erosion? You can try to figure out other questions that might be important to answer and that would help us decide what research to conduct. However, most of what we do in this research will not involve testing any hypotheses, only describing the quantitative and qualitative attributes of the systems comprising the area of investigation.

Focus on the stream for a moment. Imagine that our descriptive studies provide data that allow us to conclude the water quality is very good, there are diverse invertebrate fauna, with few pollution tolerant species, trout live and spawn in the stream, and the stream is part of a recharge system for an aquifer that may be used as a well field to supply the housing development with potable water. However, when all the other data are analyzed and interpreted, we conclude from our environmental impact study that the housing development will not damage the area's ecological systems, including the stream.

Such an environmental impact assessment involves some inferences, namely we are asking if the housing development is likely to disturb the ecological systems in the area. Thus, it is not purely a descriptive study, and yet we are not posing any hypotheses. This kind of study is not purely descriptive, nor is it a hypothesis-driven inferential study. On our continuum it is somewhere in between. Interestingly, such a study might be one phase within a much larger program of research that may eventually involve hypothesis testing.

Now, jump forward a few years after the housing development is in place. We might be asked to compare the current stream water quality, habitats, and fauna to those found during the pre-development study. For example, perhaps someone living near the new housing development no longer catches trout in the stream and believes the housing development has altered the stream quality and destroyed trout habitat and spawning areas. This idea that the stream has changed can be formalized into a hypothesis and our research would then be driven by an explicit comparison of the pre-development stream to the post-development stream. This new research would be an inferential study guided by a hypothesis about the relationships among phenomenon, organisms, places, and behaviors. The hypothesis is tested by drawing a sample from one or several populations, making measurements on these samples, and then using the sample measurements to infer whether or not the hypothesis is rejected or not rejected. Our research question for the post-development research might be the following:

Is the stream quality in the post-development study equivalent to that in the pre-development study?

This overarching question might then be formalized into one to several hypotheses, such as those listed below.
- H1: Stream water quality in the post-development study is not different from water quality in the pre-development study.
- H2: Stream habitat structure in the post-development study is not different from the habitat structure in the pre-development study.

You get the idea. But also note that even these so-called hypotheses are a bit vague. What do we mean by water quality or habitat structure? Do we need to list specific chemical species that we would study, such as concentrations of dissolved oxygen, nitrate, specific conductivity, pH, alkalinity, suspended inorganic and organic particulate matter? And do we need to be very specific about different types of stream habitats, like the distributions of gravel substrates, riffle areas, and other physical attributes, as well as numbers and kinds of organisms found in different habitats? Well, the answer is "yes." But for this discussion, we are trying to get you to think in broader terms about what constitutes descriptive studies and what constitutes inferential studies. So, let's not bother with all the details right now.

Also, you might have noticed that the hypotheses are worded as a statement that there is "no difference" between the two groups being compared. If you have taken statistics, you may be familiar with this format for the statement of the null hypothesis. You must pay very careful attention to how you word the null hypothesis. Remember, that you have created a model for some part of the real world. Your model allows you to make certain predictions and you want to test these predictions to see if they are accurate. In quantitative studies, you generally use statistics to provide an analytical framework for analyzing the data. That is, the type of predictions you are testing dictates a particular research design and the design dictates the statistical framework you will use. However, all statistical frameworks allow you only to reject a hypothesis with a certain probability of being wrong. You want this probability to be very small.

To establish a decision-making pathway, you set up a null hypothesis and an alternative hypothesis. Rejection of the null provides support for the alternative. In the ideal situation, you try to explore your model and make predictions from the model in such a way that only one of two possibilities is true. One of these is the null hypothesis, which always states that there is no difference or relationship. The other is the alternative hypothesis, which is the prediction you actually believe might be true about the real world. Rejecting the null then provides support for your alternative hypothesis. Importantly, nothing is ever "proved." We only can reject or not reject the null, and if we reject the null then we are left with the alternative hypothesis as an explanation, always with a probability of being wrong.

An interesting case occurs when we cannot reject the null hypothesis. Again, this does not "prove" the null is an accurate representation of reality, only that with a small probability of being wrong we can not throw out the null hypothesis. If we study the system again and again and again, and we never get data that allow us to throw out the null, chances are that the null statement at least partially reflects something in the real world.

So in summary, descriptive studies provide information that allows us to describe small parts of the real world. Inferential studies also allow us to describe parts of the real world but also include a process of making comparisons. These comparisons allow us to set up decision pathways that can be very powerful. Let's explore the power of the decision pathways inherent in inferential studies.

- We observe the real world and come up with questions about what is going on in the systems or organisms we are observing.
- We use the questions to construct a model of the real world and for each part of the model we specify predictions that must be true if the model is accurate.
- These predictions are refined into hypotheses in a way that rejection of the null hypothesis provides support for the alternative hypothesis.
- The predictions and their respective hypothesis structure allow us to design studies to determine if the null hypothesis can be rejected.
- Rejection of null hypotheses lead us to look at the alternative hypotheses, to then refine these and set up new predictions, which in turn allow us to set up new null and alternative hypotheses.
- We proceed systematically through tests of hypotheses, with rejections of hypotheses leading to more refined models and more refined predictions.
- As we accumulate more and more refined predictions, we begin to see patterns that fit together as general laws or principles.
- The general laws and principles become building blocks in exiting or new theories that represent more comprehensive models of the world and its various living and nonliving systems.
- These theories are refined or rejected as continued experimental study evaluates the accuracy of predictions that are necessarily a part of the theory structure.

Of course, in casual discussions about science you often here the term "fact." In the context of statistics, how do we define a fact? Well, in pretty simplistic terms a fact can be defined as something we conclude based on a very, very, very small probability of being wrong. In the language used above, a fact is a prediction that study after study has not been able to overturn (i.e., reject). A simple example is that under certain temperature and pressure combinations water will exist as a liquid, at other temperature-pressure combinations it exists as a solid, and at still others it becomes a gas. We accept the relationships between these temperature-pressure combinations and water's state as facts because enormous numbers of studies on how water changes states led us to conclude, with a very low probability of being wrong, that we could predict when water was a liquid, solid, or gas.

A slightly more complicated example is that carbon monoxide will displace oxygen on your hemoglobin molecules, so you can become oxygen starved. But the uptake of carbon monoxide molecules by hemoglobin molecules depends on the relative concentrations of oxygen and carbon monoxide in your lungs. So walking through a plume of carbon monoxide may introduce a sudden dramatic increase in carbon monoxide molecules in your lungs, immediately followed by a rapid increase in relative concentrations of oxygen as you pass out of the carbon monoxide plume. When you place a physiological system within an environmental system and both are changing through space and time, you get very complex interactions. While it is certainly a fact that breathing a certain level of carbon monoxide will kill you, the lethality of carbon monoxide is a complex set of interaction leading to tissue death from decreased oxygen transport and the destruction of critical systems (like your brain).

So, we have argued that science is a way of knowing about the real world. There are other ways of knowing, and these have their own characteristics, which distinguish them from science and from each other. For example, think about the organizations of colleges and universities. There are usually clusters of departments grouped under sciences, humanities, social sciences, and the arts. If you think about your experiences in the different courses you have taken, you can begin to outline the similarities in science courses you took and similarities in humanities courses you have taken. With a little effort, you could also contrast and compare the ways of knowing that you were taught in the sciences and in the humanities.

The biggest differences between *ways of knowing* are in the processes by which knowledge is created. Goldstein and Goldstein (1978) would say that scientific knowledge is created by a process that has the three characteristics listed above. Sit down with your classmates or faculty sometime and discuss the characteristics of other ways of knowing. This is not just an idle exercise, because Americans are not very scientifically literate and often confuse the criteria for different ways of knowing. In our technologically oriented democracy, we need a voting public with high levels of scientific literacy. Without such literacy, policy is shaped by uninformed opinion about the role of science and technology in our everyday lives.

As you will see in **VRX**, there are a number of complex problems like landfills, transportation and storage of nuclear wastes, and power generation that are absolutely integral to the way most Americans choose to live. We need to know how to use science and other *ways of knowing* to make decisions about our lives in an ever increasingly complex world. The research simulations will require you to learn science by doing science but also to evaluate the strength of your conclusions. For example, are you so sure about your conclusions that you would *bet your life* on their validity? What probability of being wrong would you use if your life depended on your conclusions, or if the lives of others depended on your conclusions? Think about this issue of probability the next time someone tells you the "facts" about something.

Research Worlds and Resource Materials

To help students become scientists and learn science by doing science, we created virtual worlds that allow students to practice the conceptual looping of the scientific method. A real problem in teaching science by doing science is that interesting research settings are often hard to get to (as well as being dangerous), equipment and logistics are costly, and students have to know a fair amount to do meaningful work. **VRX** uses virtual worlds to get around these problems.

Within **VRX**, a Virtual Office provides a gateway to the research worlds and resource materials. **Figure 1.5** shows the Office, which is the first screen you enter in the software. Remember that there are three virtual worlds: (1) a landfill, (2) Yucca Mountain, a proposed site for a nuclear waste repository, and (3) a virtual environment that provides access to coal formation simulations and a power plant model. These three virtual environments each contain four research simulations. And each research simulation has a research problem that can be used as a laboratory exercise, homework assignment, or demonstration during class. The Office also provides access to resource materials in the form of electronic books.

Basically, the office is your workspace. When you get there, you will see a desk with a laptop computer, a bookshelf, and the gateway to the virtual research worlds. The computer provides a way for you to sign in and let the system know the simulations you want to view. The *Bookshelf* is the gateway to what we call the **VRX** Reference Materials. In the *Bookshelf*, you will find a variety of electronic books. When you place the cursor over a book, it tips towards you, and the title can be read on a screen at the top of the *Bookshelf*. To read the text, you simply click on it. The book opens (**Figure 1.6**) and fills your computer screen. To access additional books, click on the colored bar at the top of the *Bookshelf*.

The electronic books are provided to support your research efforts in the virtual worlds and as reading materials that can be used to supplement readings assigned in your environmental sciences courses. Even experienced scientists often find themselves on a field or laboratory project and wish they had a reference in hand from the library or their office. You can move back and forth from the research worlds to your own personal library in the Office. Of course, you can simply go to your Virtual Office and read the materials in the *Bookshelf*. One of the most useful books is the volume tilted **VR-Excursions,** which you can use to find out more about the structure of **VRX** and the content of the various research exercises in the simulation worlds.

As mentioned, you also enter the research environments from the Office. To the right of your desk there is a simulation selector . Each research environment is contained in a chip on the simulation selector. Click on a chip, hold, and drag it to the reader at the top of the kiosk. This opens a door into the virtualworld. Take a look at the descriptions of the settings in **Study Note 1.1**, and in the **VR-Excursions** book in the *Bookshelf*.

12   Chapter 1: To the Student

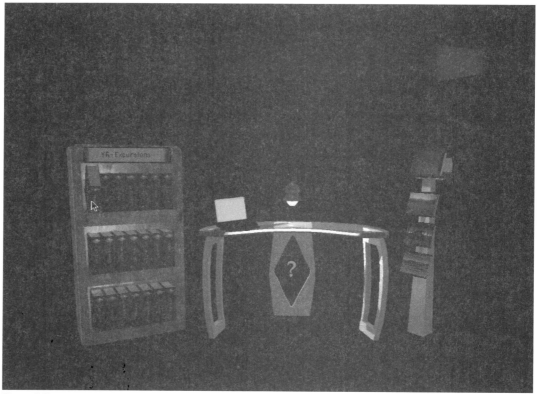

Figure 1.5

Figure 1.6

Study Note 1.1: Description of the virtual research worlds.

Land Full of Landfills – Where Has All the Garbage Gone?
In the virtual landfill, there are four research problems that can be studied, each with its own simulation. You can analyze movement of groundwater through an aquifer beneath a landfill, study the chemical composition of this groundwater, evaluate the behavior of contaminant plumes in water moving beneath a landfill, and follow gas migration from the landfill into soil and fault systems around the landfill. Inside these simulations, you move across the surface of the landfill and can sample monitoring wells. These wells are placed throughout the landfill and the area around the landfill. They are drilled in such a way that they provide access to contents in the deeper layers in and around the landfill. All of the simulations use the same virtual landfill, but different kinds of instruments and methods are used in the different simulations. The data obtained from the virtual landfill are based on real-world data, and the instruments you use in the simulations provide readings that mimic the kinds of data you would collect at a landfill.

Yucca Mountain – To Store or Not To Store Nuclear Wastes
The Yucca Mountain simulations provide a chance to study a site that has been proposed as a storage area for nuclear waste materials. Yucca Mountain is a real place. The **VRX** simulations allow you to explore the site and examine the geology of the region in which Yucca Mountain is located. From these explorations you can make conclusions about the risks involved in storing nuclear wastes at the site. You can also study risks associated with transporting nuclear wastes to the site. We have provided four venues, each with a different research focus. You can study potential volcanic and earthquake hazards, as well as potentials for groundwater entering the site or infiltration of water from above the site. As you become more familiar with the site, you will be able to calculate the risks of various problems, and compare these to risk levels established by scientists working on the Yucca Mountain Project.

Mire to Fire – Coal Power
There are three simulations in this virtual worlds. One takes you back to a Late Cretaceous Mire, where you can study processes leading to coal formation. In the second simulation, you come back to the present and study the geology of coal beds and patterns of land use in order to site a coal-fired power plant. The third simulation allows you to explore the combustion products and emissions of a coal-fired plant. Thus, you can study the process of coal formation, mining, and use that spans a period of over 70 million years. In these research studies, you will be able to explore why coal contains materials that can become atmospheric substances contributing to the formation of acid rain. You will also be able to study how to optimize power plant function to reduce emissions of these causal agents.

WHAT'S THE QUESTION?

- How the Research Simulations Are Organized
- Research Based on Interesting Questions
- Use of the Bookshelf in **VRX**

How the Research Simulations Are Organized

Earlier, we said scientists ask questions about the world, and then try to find answers to these questions. The **VRX** research simulations are organized around basic questions relevant to the different research settings. However, our students taught us an important point. What's relevant to us may not be relevant to them. As faculty, we may think we are asking interesting questions, but our perspectives are usually biased by our own interests, experiences, and research activities.

For **VRX**, we take a different approach. We feel that unless you find some aspect of the research relevant to your life, you are not likely to learn the material in a way that allows you to retain and use it. To be functional citizens in a technology-oriented democracy, you will have to make decisions as voters about an increasing number of environmental policies and regulations. For example, in the very near future, there will be very serious discussions about moving nuclear wastes across the country to storage sites like that proposed at Yucca Mountain. Do you want these wastes moving through your city? If not, why not? How will you make sensible decisions about how nuclear wastes should be regulated, moved, and stored? How would you vote on these issues?

So, we developed laboratory exercises for each research simulation in each of the three virtual worlds. Then, we grouped these laboratories into chapters. Each chapter has four sections. We boldy call the first section, **"Who Cares?"**. This section contains questions meant to pique your interest or provoke you to think about ways a particular research setting is important in your life. The second section is, **"What Do You See in the Real World?"**

This is a brief introduction to the research area and the possibilities for studying complex systems in the research setting. The third section of each chapter is called, "Dealing with Complexity." In this section, we provide the research exercises for each simulation. Finally, we close each chapter with a section called, "So, What's the Point?" In this section, we ask you to push yourself and extend your learning beyond the realm of the specific research problems you conducted. This final section has some self-assessment tools, including a case study that will challenge you to take what you learned exploring a virtual world and apply your knowledge to similar systems.

Research Based on Interesting Questions

From talking with our students, we know that at the beginning of courses it is often true that many of the questions that interest us do not interest them. No doubt, you have had courses that didn't really grab your attention at first, even though they covered areas that interested you or were required for your major. When you get to the **"Who Cares?"** section of each laboratory, it will be important for you to look hard at the questions we list, and search for reasons that at least one of these questions is important to your life. You may be able to ask related questions that we don't list. The most important point is that you find a coherent rationale for starting a research simulation.

If the questions we list don't intrigue you and if you can't find a related question that interests you, please take a few minutes and enter the simulated world. Go through it, and try some of the research activities embedded in the virtual environment. Ask yourself how the virtual environment is constructed, in what ways it is real and in what ways it simplifies the real world. Then return to the questions in **"Who Cares?"** and also read the other sections in the simulation you are studying. Loop through the conceptual framework we are proposing, and find out how and why it is related to the course you are taking. In other words, start the process shown in **Figure 1.4**, conceptually looping through the process of studying a problem, and finding areas that interest you and allow you to become intrigued enough to generate and then try to answer a question that interests you.

This approach is fundamentally different from what you may have encountered in many science courses. We are urging you to act like scientists and develop a process of studying the world. We show this process in the simplified version within **Figure 1.4**. Those of us who do research as part

---

Study Note 1.2: Why do we read so much?

Let's take an example. Suppose we want to know if material is seeping from a landfill and polluting an aquifer that may supply the wells of people living quite a distance from the landfill area. Also, suppose that our interests are principally focused on the organic compound benzene, which can cause a number of problems in human physiology. A reasonable place to start would be a literature search that used "benzene" as a keyword. Try this yourself. Go to your library, or get on the Internet, and do a search on just "benzene." You will turn up hundreds of references. To narrow your search, you might look for references on benzene in landfills, soil chemistry of benzene, transport of benzene through soils, benzene as a water pollutant, measuring benzene in soils and water, and toxicity of benzene to humans. Each of these areas is important to what you want to know, namely, is benzene seeping from a landfill and polluting the wells from which people are obtaining water.

You might also begin talking with colleagues or other scientists who may have worked on this problem or similar problems. In short, you begin looping through the information available on how, when, and why benzene might pollute water that is eventually used by humans. Even with good focus and considerable knowledge about benzene, you will read a lot of papers and books that provide general background, but do not specifically help you develop a study that can answer your question. With luck, there will be some papers that have addressed the question you are interested in, and you can use these to guide your own studies. On the other hand, you might find that none of the literature you read or the scientists you contact have the information you need. Or, you may decide you want to find a new approach to answering the question.

All the background work you did was still important. In order to save time, money, and effort, you really had to know what work on benzene pollution had already been done. It doesn't make sense to re-invent the wheel, or in this case re-invent the analysis of benzene and methods to study its movement into aquifers. Yet, it is always possible to improve, refine, and expand our knowledge in an area. So, push yourself to find questions that interest you. Just don't forget you may have to read and sort through a fair amount of previous work in order to ask and then answer questions that engage you.

of our daily lives loop through many different conceptual frameworks looking for interesting questions. As part of this ongoing process, we read research papers in journals and books on the subject, work in our laboratories or on field research projects, and talk with colleagues. One point many of our students do not realize is that much of what we do is preparation for understanding how to ask a question and then trying to answer it. We often find that much of our background reading is not specifically focused on the question we are trying to ask and answer, but provides a broader context for the area we are studying. Take a moment to review the example in **Study Note 1.2.**

Use of the Bookshelf in VRX

When we were students, and even now in our more advanced age as scientists and teachers, we have often wanted to have all of our books and papers around when we were working on research projects. This is usually possible when we are working in labs next to our offices, but not so easy when we are out in the field, miles or thousands of miles from our office or university library. In the information age, telecommunications has made it easier to access and use a variety of information bases, even from fairly remote sites. In **VRX**, we mimic this access by incorporating a bookshelf in the Virtual Office.

Earlier, we described the *Bookshelf* as a set of electronic books. These books can be found to the left of the desk in the Virtual Office. You can use these electronic books any time you enter the Virtual Office. For example, check them out at the start of an exercise or by leaving a simulated research environment for a short time and looking at a reference.

Study Note 1.3: How to use the electronic books.

The electronic books have some built-in functions that make them easier to read than most regular textbooks. The left side page is regular text. However, the right side page is an **information window** that provides a lot of support for you while you are reading. For example, some of the text words are colored blue. Click on these, and a **Note Card** opens on the right page with a definition. This provides an instant electronic glossary. Important concepts appear in text marked in green. Click on these, and more detailed explanations open on the right page. A purple word will allow you to access a video clip that helps explain textual material. In other words, you can read text just as you do in a paper book, but the electronic resources of **VRX** allow you to open sources of information on the right side page.

The Bookshelf is designed to represent a transition from traditional textbooks to the kinds of information sources that are now becoming available and will be much more common in the near future. Importantly, by reading the text on the left side of the electronic books, you will get the basic information you need in the subject area covered by the book. One way to use the electronic books is to read the text on the left side quickly and get an overview of the subject. Then re-read the text, and use the color-codes to access more information. This second reading will allow you to use the **information window** to help you expand and consolidate your knowledge. In a later section, we will discuss how you can identify the ways you learn best, and then develop a strategy to effectively use the Bookshelf.

You will note that the *Bookshelf* is not really like the ones found in most libraries, nor are the books exactly like hardcopy texts. The books are electronic in the sense that their text, figures, and other visual aides are all contained within the software and can be read from your computer screen. When you open them, you will find fairly traditional text on the left hand side and an **information window** on the right hand side. The text on the left side reads very much like any hardcopy textbook, with a few interesting exceptions. You will note that some words are color-coded. When you click on these words, the **information window** on the right side provides additional text, figures, or video clips. The **information window** is an electronic resource tied to textual material on the left page. In **Study Note 1.3**, we provide some guidance about how the electronic books work.

Who's Responsible for Teaching and Learning?
- You Are, You Are, You Are!
- Making Connections
- Developing a Context
- Learning by Doing

You Are, You Are, You Are!

VRX is, in part, the result of lessons we learned from our students. Over the years, our work with thousands of students has taught us a great deal, and sometimes the lessons were not all that easy to comprehend. We learned what students liked, and what they absolutely hated. We learned how students are very diverse in their expectations for a course and success in a course, as well as how diverse they were in the value they placed on their success in learning course material. In some of our earliest work with students, we tried to take them out into the real world, giving them a chance to work in complex environments and interesting places. It became clear that students actually learned a lot of science by doing science. Furthermore, they did not have to know a great deal about a subject area when they started. After all, who takes a course that covers material you already understand and can apply (although this does increase your GPA). This idea of **learning science by doing science** was a very powerful lesson for us.

The limitations of doing high quality science with students also became painfully obvious. If you have a 2-3 hour laboratory, how far can you travel to a research site without using up too much laboratory time? How much equipment for student research can you buy with limited budgets? How do you take students to really complex and interesting, but sometimes dangerous places? Again, students were our teachers, because they helped us bring the world into their lives by using virtual settings that offered interesting and complex research opportunities.

Now, the point we want to make is that you are ultimately responsible for learning. Sure, faculty can talk at you, talk to you, work with you, assign this and that for you to do, but you have to decide why it's important to learn the material. Once you decide, there is still the big part of learning, namely getting down to business and working through the simulations. Just try to keep in mind that success in learning is shaped by many factors, including the value you place on the learning process and your expectations for success. Your expectations and values are shaped strongly by socializers in your life, like family and friends, but also the faculty teaching you. It will be very important for you to look closely at the value you place in studying the material in **VRX** and your expectations for success in learning the material. Part of this self-exploration will require you to make meaningful connections between the conceptual frameworks of environmental sciences and their relevance to your life.

Making Connections

We feel that an important first step is to figure out some connections between the content areas you are studying and your own life. It is certainly true that our education system often imposes course requirements that do not always fall within your interests. In such cases, you may be hard pressed to figure out meaningful connections between course material and your life, other than that you have to take the course to graduate. However, the **VRX** research settings introduce you to problems that have very real, everyday impacts on your life. Your quality of life, and the quality of life of all people on earth, depends on solving major environmental problems related to solid waste, nuclear waste, and use of fossil fuels.

The responsibilities of citizens in a technology-oriented democracy like the United States include being scientifically literate enough to make good decisions about environmental issues. In simple terms, during your life you will be asked to vote and so shape decisions about how local, state, and federal governments respond to environmental problems and concerns. You will need at least a basic knowledge of environmental sciences in order to figure out the consequences of different responses by government agencies. Unfortunately, the sciences are often taught in ways that treat students as passive learners. One really important consequence of such passive learning is that students seldom get any experience with how scientific knowledge is actually created. Without such knowledge, people are ill-prepared to evaluate the quality of scientific evidence. If you don't know the quality of the scientific information, it is difficult to make meaningful decisions using such information.

It is also notable that most science courses don't examine the possible outcomes of making decisions based on the data from a study. Of course, in the real world, decisions are not made by only using scientific information. While **VRX** was designed to give you a very realistic sense of how to do science, we also add elements that involve students in thinking about historical, political, aesthetic, and cost-benefit factors that shape decision-making in real-world situations.

Study Note 1.4: Asking the questions that connect you to a course.

Why? How? What? These are questions about your course that we want you to ask all of the time. **Why** are we studying these facets of the environment? **Why** are they important? **What** are the most important concepts? **How** do we study for this course? **Why** is each assignment worth a percentage of our final grade? **Why** did the instructor choose this topic to present? **How** much time should I spend studying this area? **Why** do we complete a particular laboratory in the fourth week rather than the seventh week of the term? **Why? How? What?**

VRX can be used in many different ways, with quite an array of courses. If the simulations are assigned as part of a particular course, your instructor will undoubtedly explain how and why to use the simulations, which simulations to enter, and what kinds of research to conduct. We encourage you to think about why faculty assign the **VRX** simulations and how these assignments contribute to your learning of the course materials. Often, the course syllabus will describe the purpose for using **VRX** and other laboratory materials. Keep pressing yourself to really understand the purpose that working on the simulations will serve in your course. When in doubt, ask the instructor. If you don't understand how and why the **VRX** simulations are being used, it's a sure bet that many of your classmates also do not understand. Try answering the questions in **Study Note 1.4**.

It is also important to understand that faculty try, with varying degrees of success, to get to know their students. It is often hard for a faculty member to know everyone by name, particularly when teaching classes with enrollments over 40, as well as teaching several classes each term. It is pretty much impossible with large classes for the faculty member to meet with each student individually, get to know them well, and explain how and why a course is taught in a particular way. The course syllabus and a few introductory lectures generally provide the framework for a course, and may provide the only guidance given to students about the course format, expectations, grading, and assignments.

VRX can do a lot to help you learn, but you have to know "how" and "why" it is being used. Otherwise, you will not have a context for learning, and so not be able to find the meaningful and interesting questions that will motivate your learning and mastery of the course material. If you feel that your instructor does not make an explicit connection between the use of **VRX** and the course content, ask your classmates if they see a connection. If most of your classmates do not see a connection, then the connection was probably not well developed by your instructor. It is then very reasonable for a group of you to ask for clarification during or after class, or during the faculty member's office hours. Take a look at **Study Note 1.5** for some advice about how to approach faculty members.

Study Note 1.5: Some advice about approaching a faculty member.

Students are often hesitant to approach a faculty member. Our students have told us they don't want to appear stupid, they don't have a clear question, they're too nervous to talk to the professor, or they think that the faculty member doesn't want to be bothered. It has been our experience that if one student has a question, there are always several to many others who have the same question. Use your classmates as a resource. Find out if they know how to find the information that will answer your question, such as a chapter in your textbook or notes taken during class. Start a study group of classmates with the same level of motivation as you. Use the group as a discussion forum to identify questions that none of you can answer, and take these questions to your instructor. Go as a group if you are nervous.

Remember! How? Why? What? Keep trying to find out what you are expected to learn, how the instructor wants you to learn, and why the material is important to you, to your major, and to the instructor. Within these questions, try to understand the role of **VRX** in the course. Why are you using **VRX** research problems? What do they contribute to your understanding of the course material? What assignments will you have to turn in from the **VRX** research? How much will your work on the **VRX** problems count in your course grade?

Developing a Context

So why are the environmental sciences relevant to your life? Try to answer the questions in **Study Note 1.6**. If you don't know how to answer these questions, you don't know enough to vote on related issues in a local, state, or federal elections. Furthermore, if you don't know how to vote on environmental issues in an election, you may allow degradation of the environment in which you and your family live. Don't just contribute passively to the degradation of global systems. Think deeply and broadly. Make decisions based on objective criteria.

> **Study Note 1.6: Do You Know The Answers To These Questions?**
>
> - Where does your water come from, and how do you know if it is safe to drink?
> - If you didn't have your trash and garbage carried away, how long would it take to fill up your house?
> - What site is used to hold all of your garbage and your neighbors' garbage? Is it a landfill that meets regulations for protecting the surrounding environment?
> - What countries produce the most solid waste per individual?
> - What countries produce the largest amounts of hazardous materials per individual?
> - How much electricity do you use daily, and what energy sources are used to generate that electricity?
> - Take stock of a day in your life. Can you account for all of the potential pollutants generated in the production and distribution of your food, the purification of the water you drink, the processing of your sewage, and the disposal of the trash and garbage that you discard (even that paper cup from your favorite espresso shop). In the production and expenditure of the energy you use for cooking and keeping warm or cool? In the creation of the clothes you wear, the cosmetics you rely on, your recreational equipment (how were those cross-training shoes made?), the educational materials you are using (even this book and the VRX software)? In the manufacture, maintainance, and operation your car? In all activities necessary to provide public transportation by plane, bus, train or subway?
> - What would happen to the world if everyone on the planet lived the way you choose to live? Would we all be able to live within a sustainable environment?
> - Is nuclear power a better alternative than fossil fuel power in terms of short- and long-term pollution of the earth?
> - Why do we care where a landfill is sited? A hazardous waste storage facility? A nuclear waste depository?

Learning by Doing

If you can find any relevance in at least one of the questions listed in **Study Note 1.6**, you can begin to make learning about the environment more personal. This means you can find some value in studying environmental sciences and using **VRX** to enhance your learning. The key to using the simulations in **VRX** is *doing*. That is, we want you to **learn** science by **doing** science, and the research simulations are a good place to start. While there are many interesting undergraduate laboratories that offer some experience in research, most have a limited scope, because research can be both costly and logistically difficult (e.g., travel to a research site, amount of time needed to conduct a study, the need for special equipment, and so on). Furthermore, you usually have to have some knowledge about a system before you can study it, at least in any meaningful way.

We mentioned in **Introduction to VRX** that the materials presented in Chapters 3-5 give some clues about what prior knowledge you need to begin a research simulation. We also described the *Bookshelf* and its electronic books in the Virtual Office. You can decide to read some portions of the books prior to entering a virtual research setting, or establish a study schedule in which you go to the Bookshelf and spend time reading the electronic books, just as you would go to a library for reading on reserve. Please note you can easily go in and out of the simulations to consult the *Bookshelf.* We want to keep emphasizing, however, that it is critical for you to connect your work in the simulations to the material covered in the courses you are taking.

CHOOSE A PATHWAY TO LEARNING!
- Take Charge of Your Own Learning
- You Got Game
- Work with Your Instructor

Take Charge of Your Own Learning

We can't kid you about the fact that if you want to be good at something, you have to work, and then work a little more. One very striking feature of our lives as faculty is that in every college and university at which we have taught, in every one of our classes, the vast majority of the students were capable of successfully completing the work with a decent grade. Now, of course, many did quite well, but a distressing number did not do the work they were capable of doing. We take some responsibility as their teachers, but not all of the responsibility. Sometimes, students just didn't care about working with us to learn the material (although they did care about the low grades we passed out).

VRX is one of our efforts to change the nature of education and provide more flexibility while also integrating deeper critical thinking to undergraduate science courses. We think the software does a pretty good job of offering you opportunities to learn science by doing science. If you don't *show up*, the software is pretty much useless, no matter what its potential as a set of **mindtools** designed to help you learn areas of the environmental sciences; and by *show up*, we do not mean simply turning on the computer and drifting through the virtual research settings or casually clicking through the electronic books.

The exercises for the simulations in Chapters 3-5 of this manual are a framework for research. Each has research questions and some basic guidelines to follow as you conduct the research. Your instructor may modify the research questions slightly or use the simulations to conduct research not described in Chapters 3-5. Our point is that you can use the virtual research settings and the *Bookshelf* to learn a lot about the real world. However, the type of learning designed into **VRX** is **student-active learning**. You have to be an active participant. The research settings are not places that foster learning by simple memorization. You have to construct knowledge in much the same way that scientists construct the body of knowledge we call science.

VRX uses what we call the **Triple E** approach, developed originally by researchers who studied science education and promoted the idea of **learning cycles**. The **Triple E** approach is outlined in **Study Note 1.7**.

You Got Game

Let's continue with a short discussion of ways and means for you to take charge of your own learning. You want to be a good student? Put some time in studying. How long, you ask? Well that depends on you. Some people have tremendous abilities to read fast, comprehend quickly, and understand patterns. Others need a little more time to think things through. Here are some rough guidelines. Remembering the thousands of students we have worked with, our general impression is the good students were good because they studied the material we assigned. This meant they were spending at least an hour every day on each subject. Certainly, some subjects require a little more time each day, and some may require a little less, but not much less. So, putting in your time is critical. Then, You got game!

> **Study Note 1.7: The Triple E approach.**
>
> - Engage in the subject.
> - Explore the subject domain.
> - Extend your knowledge to related systems.
>
> To **Engage Yourself**: It is important to identify the research question you are trying to answer and to identify why this question is relevant to you and is, therefore, worthwhile to spend time studying. You remember that earlier, we said each virtual research setting is described in a separate chapter, and each of these chapters has four sections. In the first section of each chapter
>
> there is a section called, "**Who Cares?**". In this section, there are questions meant to engage you by intriguing you or provoking you to think about ways a particular research setting is important in your life.
>
> **To Explore the Subject Domain:** The second and third sections of each chapter help you enter the research setting and provide some direction towards research areas. We call the second section, "**What Do You See in the Real World?**". In this section, you will find a brief introduction to the research area and ways to study complex systems in the research setting. The third section of each chapter is called, "**Dealing with Complexity**". This is the section in which you will find specific research exercises for each research simulation.
>
> **To Extend Your Knowledge to Related Systems:** We have a fourth section in each chapter called, "**So, What's the Point?**" This section provides some self-assessment tools, including case studies that will challenge you to take what you learned conducting research in a virtual research setting and apply your knowledge to similar systems.

We tried to make **VRX** entertaining as well as educational. However, the educational part can only become accessible if you put in some time exploring the virtual research settings and doing some research in these settings, as well as data analysis and interpretation after leaving the settings. If you want to play in the big game of life, you will have to work developing both qualitative and quantitative reasoning skills. These skills require practice, time, and guidance. Work slowly but surely. Don't cram for exams or leave papers or reports until the night before they are due. You may actually be able to get acceptable grades by relying on this kind of last minute effort, but it is not likely that you will retain the knowledge for very long. Thus, in the subsequent classes in your major or in a career after your university days, you may find you don't have a functional knowledge of the subject areas you studied. Find what works for you, but pace yourself in a way so that you are studying most of your courses every day, or at least every other day. Then, You got game!

However, don't try to learn everything on your own. Find or form a study group. Let's be clear about how study groups work. You can help each other learn material, even share tasks for taking, collating, and improving notes from lecture, laboratory, or reading. You can quiz each other, talk over areas which are difficult to understand, create study guides for exams, and read and edit each others' papers. However, unless your instructor assigns you a specific collaborative project, you have to do graded assignments by yourself. That is, write your own papers and lab reports, take your own tests, and so on. This may seem obvious, except we know from experience that when students work together, there is a convergence in their thinking. In **Study Note 1.8**, we offer some ideas about how to work in a study group, but hand in your own work.

> **Study Note 1.8: Study together, and turn in your own work.**
>
> Sometimes, all of the members of a study group will turn in their assignments or tests, and they will look exactly alike. A simple example is that if a study group decides on a definition for a term, then everyone basically memorizes that term in the way the group defined it. Consequently, everyone will get it right on a test if the definition is fundamentally sound, or everyone in the group will miss it on a test if the definition is inappropriate. On an examination, a faculty member would expect students to try to find and use the correct definition, and the writing of the definition will be similar among the students who know the definition. Problems can arise in papers and lab reports that are supposed to be completed individually. A study group decides on definitions for terms, how to write sections of a paper or lab report (such as a Materials and Methods section or a Results section), and then everyone in the group uses nearly identical language in their papers. When an instructor reads these papers, it will look as if someone copied from someone else, when in fact it was a group effort that was individually expressed in nearly identical ways.
>
> We strongly advocate study groups, and yet, as faculty, we know that sometimes people don't do their own work and borrow heavily from others. Thus, we encourage you to form study groups under the guidance of your instructor. Let her or him know that you are studying in a group, and seek their advice about how you can take advantage of group studying, but still hand in individualized work that meets the faculty member's standards. The **VRX** research simulations can be done by small groups, usually 2-4 students. Working together has a number of advantages, including navigating through the virtual environments, collecting and recording data collection, monitoring each other's data collection, questioning choices about how to proceed with the research, and reminding each other about the research question and the goals of the research. Then, *You got game, he got game, she got game, we all got game.*

Whether you study alone or in a group, it is crucial for you to regularly assess your understanding of the material you are studying, and we don't mean waiting for an exam. A few paragraphs earlier, we said to study every course on a regular schedule, at the very least touching each subject every other day but preferably every day. At the end of each week, try to test yourself with questions or problems. You can use a study guide if there is one recommended for the course. Look in your text and see if there are questions or problems you can use. Obtain practice examinations from your instructor, or try re-entering **VRX** simulations and thinking through how the research relates to other course work.

As we write this students' manual for **VRX**, we have spent time thinking about our own experiences and the experiences of our students after we finished our doctoral work and entered academia as faculty. It is important for you to know that even though we are successful faculty and were pretty good students, each of us had courses that were really difficult, and in which we had to struggle to master the subject areas. We are asking you to do something that we have done, and we are asking you to do something that we have asked our many students to do. Just give it a try!

Work with Your Instructor

What a novel idea! Of course, many of you have contacted your instructors during your academic careers. We hope all of you will make this a regular practice in every course. It is important to get to know your instructors and for them to get to know you. While there are faculty who are excellent teachers and faculty who not very accomplished teachers, the vast majority are dedicated to teaching their courses as best they can. As part of this effort, they should be willing to explain why they chose to teach a course in a particular format and to cover a particular set of content domains. In fact, most faculty usually try to address these issues in the first few class periods as well as in their syllabi.

In regards to **VRX**, you will probably be asked to use the software as part of a laboratory or other course assignment. In order to learn effectively, and, as a consequence, get a reasonable grade, you have to find out what you are supposed to be doing and the scope of the assignment. You can try to find answers to questions you may have by reading this manual and using the software. However, if you still have questions, it is critical to talk to other students or to your professor.

Most universities and colleges require that the faculty be available for work with students during fixed office hours, and the vast majority of faculty post these hours on their door or have them posted in their departmental office. As we said before, however, a faculty member may have 2-5 classes each semester, and each class can have many people in it, sometimes over 100 people. You can figure out a faculty member's time on the job. For each 3 semester hour class, there are 3 contact hours per week and, usually, an additional 6 hours of preparation for that class, including getting lectures ready, preparing materials, and grading. That's at last 9 hours of time spent on each class during each week, so if an instructor taught 4 classes that would amount to 36 hours of work.

The majority of faculty are required to help with departmental and college or university committees, which takes up another 2-4 hours per week. Add this to the class time, and your instructor has a 40 hour a week job. However, there are still the required office hours, usually 4-8 hours when instructors must be in their office and available to students. In addition, add at least 10 more hours of time to their work week if the faculty member is required to do any kind of research. This brings the faculty work week to over 50 hours. Also, take a look at the number of office hours. Say there are 4-8 office hours a week. How much time would each of 200 students get if they visited their instructor every week? **Study Note 1.9** gives some more advice about working with your instructor.

Remember, don't just drop by a faculty member's office unless they have open office hours. Like you, they have work to do other than just the work required for your class. Imagine a situation in which you have only a limited time to finish studying for a test in your mathematics class, and your environmental sciences instructor unexpectedly shows up at your dorm. You answer the knock at the door and the instructors says, "Hey, I just wanted to ask you some questions about how my lectures are going in environmental sciences." Would you be able to just stop what you're doing and talk? Well, if your mathematics test was important and you really needed to finish some critical studying, you might respond to the professor, "I'd be happy to talk with you, but could you come back during the time convenient to both of our schedules?"

In our conversations with students, we find they often don't understand that the faculty member is not available all of the time for a specific course, but is required to schedule office hours to be available to answer students' questions. Also, our faculty colleagues sometimes forget what it was like to be an undergraduate, so they may speak too abruptly or give non-verbal cues that suggest they really do not want to talk to students. In truth, there is a middle ground where faculty and students can, and should, meet. This middle ground must be based on mutual respect and an understanding that students are in school to learn and faculty are there to teach. Students and faculty must find that shared space for a safe teaching and learning environment that is equitable for both.

Study Note 1.9: More advice about working with your instructor.

Suppose you have a question about **VRX**. Your time is valuable. Your instructor's time is valuable. Working individually or in a study group, you can prepare a set of specific questions to take with you to the instructor's office. Find out when the instructor has scheduled office hours, and if you need an appointment. Often this information is listed in the syllabus. Don't go just any time. Always go during the office hours unless you have classes scheduled during every office hour your instructor offers. In this unusual case, call and make an appointment that meets both of your schedules. Explain to your instructor that you have a set of questions. It will help you and the instructor if you write these down and take a copy for the instructor. Go over each question until you understand the material. If you can't cover each question during the time your instructor has available that day, come back as soon as you can and get through your list of questions. Finally, take notes on what your instructor say, so you can refer to them later.

If you go to the instructor's office with 2-3 people from your study group, you can combine the study group's questions about **VRX** into a coherent list. All of you should take notes and ask clarifying questions during the meeting. After the meeting, spend some time debriefing and comparing notes so you are sure all of your questions have been answered, and all of you agree about the answers and now understand the **VRX** material. Going with a small group of classmates helps in the following ways: (1) you all get to hear what the instructors has to say; (2) you can compare notes to make sure everyone has the same understanding; (3) usually, each student will pick up something the others missed; and (4) faculty members can more effectively use their time to help you, because they are dealing with a group of students.

Getting Started

INTRODUCTION TO THE VRX SYSTEM

What is VRX?

Virtual Reality Excursions (**VRX**) is a set of three virtual worlds in which you can study complex environmental problems. This software was developed so that you could conduct realistic, high-quality research within realistic simulations. The virtual worlds include: (1) a landfill, (2) Yucca Mountain, which is being considered as a national repository for nuclear wastes, and (3) a virtual worlds that includes a coal formation simulation and modern settings such as a coal power plant management model. Each of these three virtual worlds has four research simulations, and each simulation has a research problem that can be used as a laboratory exercise, homework assignment, or demonstration during class.

System Requirements

Required Software: Apple Quicktime
VRX requires that Apple Computer's Quicktime software version 3.0 (or later) be installed properly. The Quicktime installer program is included on the VRX CD-ROM in the folder named Quicktime. You can also download the latest Quicktime from Apple Computer's website (www.apple.com)

Hardware and System Requirements
VRX is a hybrid CD so it runs on both Macintosh and Windows-95/NT platforms.
In order to use **VRX**, you will need at least one of the two following systems:

>An IBM-Compatible Computer, Pentium II, 150 Mhz (minimum processor speed), 32 MB (minimum) RAM, 640x480 screen size, thousands of colors, 100 MB hard drive space,* 12x CD-ROM drive, Soundblaster 16 Soundcard compatible, stereo speakers or headphones, running the Windows 95 Operating System or NT 4.0.

>An Apple Power PC 604E, 150 Mhz 150 Mhz (minimum processor speed), 32 MB (minimum) RAM, 640x480 screen size, thousands of colors, 100 MB hard drive space,* 12x CD-ROM drive, stereo speakers or headphones, running MAC OS 8.1.

Starting VRX

VRX is designed to run from the CD-ROM drive in your computer so there is no installation other than opening the CD-ROM drive, placing the **VRX** disc (Graphic side up) into the drive and closing the drive door. To launch **VRX**, find the CD-ROM drive on your computer desktop and open it using your mouse. Double click the VR-EXCURSIONS icon to start your research. **VRX** will not install anything on your hard drive, nor will it attempt to contact anyone or anything through your internet hookup. **VRX** should not be run simultaneously with programs other than the operating system. Please note that **VRX** is designed to run from the CD-ROM, so it is important to have a CD-ROM drive with a speed of at least 12x.

*Ideally, the system should have 100 *contiguous* MB of free disk space. There are comercially available desktop utility programs that can help clean up your hard drive.

Using VRX in a Computer Laboratory

The installation in the computer laboratory depends on the systems available. Your faculty member or the computer laboratory supervisor should have already installed VRX on individual machines or on a network system and should let you know where it "lives." In general, students are not allowed to install software on a laboratory computer. Please check with your instructor if you have questions about using the VRX in a computer laboratory setting.

What You Should Expect (quality of visuals, speed, common problems)

VRX uses the Apple Quicktime media layer system. This includes Quicktime Video and Quicktime VR Video which allow for good graphics and digital video. So first and foremost, the graphics should be of high quality with good color. If the movies and graphics appear blocky or just look terrible, check that your video card is set to thousands of colors. **VRX** is not designed to function at a 256 color depth. If you can't see any digital video, please check that Quicktime 3 is installed properly.

Second, the system should respond quickly and in a very smooth manner. In particular, there should be no jerky motions or unannounced long delays as you move through the virtual research settings or use research equipment. If you notice slow, jerky, or delayed software responses, it may be that your particular system requires additional RAM, your processor does not meet the basic requirements, or your hard drive is full or badly fragmented. If the QTVR movies appear banded or "breakup" you may need to find an updated video driver for the computer's video card. Please consult the manufacturer of the video card or computer for additional video drivers for your machine.

A VIRTUAL WORLD WITH REAL PROBLEMS

A Quick Tour of a Landfill

Assuming that you have started the **VRX** software and are staring at the office, you should take a quick tour of the landfill simulation. To begin you should log in to the system so that you can identify yourself and choose the appropriate simulation. Click on the computer that is sitting on the desk in the virtual office. (Click = press and immediately release the mouse button.) A screen should appear that asks you to enter your name and social security number (the SSN is optional) and allows you to choose a simulation and level (**Figure 2.1**). Go ahead and enter your name, and choose the Solid Waste Landfill Simulation - Level One. Then you can click the button that says "USE THIS INFORMATION TO LOG INTO THE SYSTEM" to continue. The computer should confirm your identity and welcome you to VRX. To get back to the office, click the office icon in the lower left corner of your screen. (Note: the office icon is the sure way back from the deepest part of the simulations.)

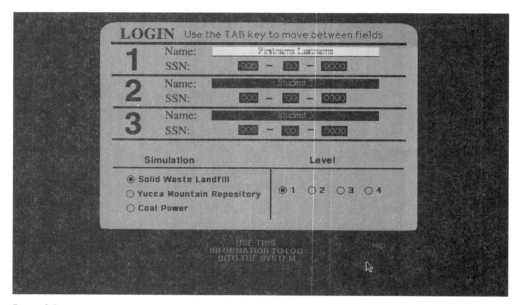

Figure 2.1

At this point you could jump right into a simulation, but wait—first find out how to use one of the instruments that you'll have access to in the landfill. Run your mouse over the books until you discover the book entitled "INSTRUMENTATION." Click on the pulled out book to open it. Using the scroll arrow scroll the left side page down until you see the information on the Sonic Water Level Probe (**Figure 2.2**). Click on the colored text that is part of this instrument's description. Once you have learned about the operation of the probe close the book and return to the office by clicking on the office icon.

OK — now look at the simulation selector that holds three colored chips over on the right side of the office. Notice that as you run the mouse cursor over those chips that they depress and a noise is heard. Click and grab the top "chip" and move it all the way up to the top of the selector's purple screen to launch the Solid Waste Landfill Simulation.

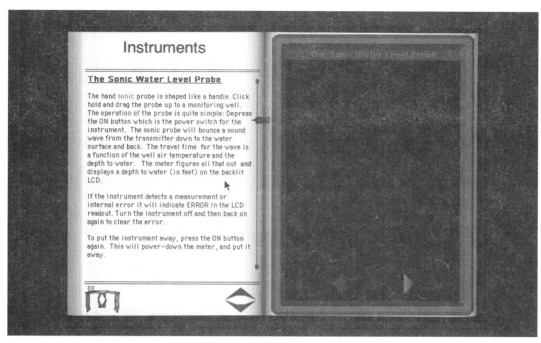

Figure 2.2

The screen in front of you should look similar to **Figure 2.3**. Click on the blinking "LANDFILL" site to enter the virtual world. You will find yourself in the landscape shown in **Figure 2.4**. Place the mouse cursor into the middle of the landfill scene. Hold down the mouse button and move the mouse to the left inside the scene — the scene should rotate around your position as you get your first look at the 3-D world. Try other directions to see how the movement works.

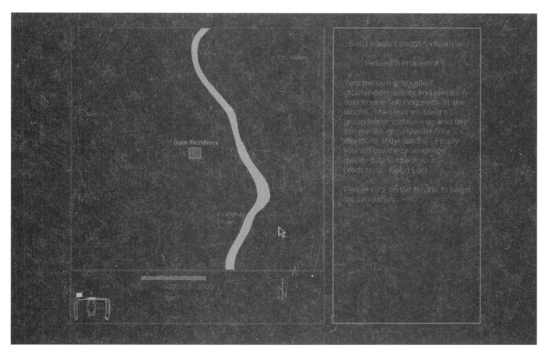

Figure 2.3

32 Chapter 2: Getting Started

Figure 2.4

Now its time to drive repeatedly. Move the scene around until you can see down the landfill road. By clicking repeatedly on the road you can move farther into the landfill world. Continue down the road until you find one of the wells. Get as close to a well as you can and double-click the well. (You may notice that the cursor changes depending on what your doing. When you are able to click down a road the cursor appears as a fat forward arrow, when you are close enough to a well to double click it the cursor appears as a hand.)

Once you have double clicked on a well, a new scene showing a close-up of the well or wells will appear. You will notice a row of instruments on the bottom of your screen. Click and drag the Sonic Water Level Probe up to one of the wellheads (the wellhead is the place where you can see the end of the pipe sticking out) . You should remember from your foray into the bookshelf just a few minutes ago what this meter does and how to operate it.

Record the datum that the meter gives you — what does this number mean? You have just taken a water level measurement in a groundwater monitoring well! Turn off the probe and click on the landfill map to get back to landfill road. Once you are back, notice the map on the right hand side of the screen — there is a green dot that indicates your position at the landfill. What next? Go ahead and wander around the landfill — but please don't eat any fish from the landfill pond! When you are ready to return to the office click the office icon on the lower left corner of the screen.

OK, that is the quick tour. Suppose you want to explore other simulations or different levels? To do that go back to the desktop computer and make the changes you want. To get out of **VRX** and back to the real world, click the EXIT sign. Some credits will start — to bypass them, click the EXIT sign again.

Land Full of Landfills – Where Has All the Garbage Gone?

WHO CARES ABOUT LANDFILLS?

Did you throw anything away this week? Where did your garbage go? Most of us have a vague idea of what happens to our household waste. We know that there are garbage trucks and landfills, but once the waste bin leaves the curb in front of our house, it's conveniently "out of sight and out of mind". In the United States, this waste usually ends up in a solid waste landfill.

Should you care? Some years ago, we designed an exercise for students that shocked us and them. We asked students to collect their solid waste over a 24-hour period. They placed all of their solid waste in a heavy duty garbage bag and brought it class. We weighed everyone's bag individually and calculated some simple descriptive statistics such as average solid waste produced by class members, the variance, and the range of solid waste produced. These statistics gave us a sense of the amount of solid waste produced and the variability among students. We also categorized types of solid waste collected. This qualitative analysis allowed us to talk about all the different kinds of things we routinely throw away and, also, about how differences in lifestyles can lead to differences in the categories of solid waste an individual may produce.

This exercise was instructive, because it brought home how much we actually throw away each day. We could do simple modeling of individual annual solid waste production as well as using average estimates to estimate solid waste production by the university community. However, we did not look at a variety of other sources of waste that are actually associated with our daily lives. For example, everything we used and even the things we threw away had to be produced. There is also considerable solid waste produced during the processes of manufacture and distribution of the products we use daily. These wastes are hidden, because we don't have to deal with them.

Take the clothes you wear. At every stage of production of your clothes, there is waste produced, and someone has to find a place to throw it away. What happens to the cloth scraps, thread, and unused dyes? Clothes then tend to be transported from the site of manufacture to distant retail sites. Along the way there is packaging you never see that gets tossed. Think about all of the paper work, like advertising, billing, transportation labels, and cardboard boxes for shipping. Get the point?

You could go through your life and inventory all of the things you use or do routinely. Then for each, try to think about the solid wastes created in the processes of manufacture, distribution, and use. How much solid waste do we produce, and where does all the solid waste go? See if you can answer these questions.

- List all of the things you threw away during the past seven days. What was the volume of all that solid waste? How much did it weigh?
- Take you favorite clothes. Figure out the manufacturing process, and estimate the waste involved in making your clothes, shipping them to the store where you brought them, all the wrappings, and the bag the sales people gave you to carry the clothes home. Where did all that waste go?
- There is a lot of talk about recycling. Do you buy materials that can be recycled? Why or why not? How much of what you buy in the course of an ordinary day is recyclable?
- How much of the material thrown away every day is hazardous to your health? How would you know?
- Where is the landfill closest to your home? Is it constructed in such a way that it will contain wastes without allowing harmful materials to seep into the surrounding environment?
- How about those CDs? Can you estimate the waste involved in producing, distributing, and selling those tunes you love so much?

Get the point? Solid waste is a big deal. It's everywhere, it's everywhere. We can choose to change our lifestyles and reduce solid waste production. However, it is unlikely we will be able to change our lifestyles to the extent that we produce no solid waste. So, we need to study how solid waste is handled and mishandled. Where it is stored, and how it is stored. What can go wrong, and how we know if our storage areas aren't containing the waste we dump there.

WHAT DO YOU SEE IN THE REAL WORLD?

Modern landfills (approved by the EPA) are expensive engineering marvels that hardly ever leak their contents. However, in the United States there are plenty of older landfills still operating that don't have "high-tech" liner systems. Many of these do leak. A few large landfills are estimated to leak up to one million gallons of leachate into groundwater each day!

While we live in the country that generates more solid waste than any other, we are also arguably the most environmentally regulated country. These regulations make relatively clear and swift pathways available for us to deal with environmental hazards such as leaking landfills. Hopefully, such regulations reduce short-term and long-term health hazards from landfills. On an international level, there are many countries with less stringent regulations. During the next decade, we will probably be hearing more and more about health hazards associated with landfills in less regulated countries

DEALING WITH COMPLEXITY

Storing the waste products of our technologically oriented society is not an easy matter. People used to just bury solid waste. Out of sight, out of mind. Unfortunately, we were out of our minds not to think through the processes of decomposition or degradation, the interaction of waste materials or their breakdown products with soils and water, and the possible movement of materials from the site of disposal through soils and groundwater to areas quite remote from the dump site. The truth is, storage of our waste products is a very complex problem. First, there are lots of different kinds of waste, each with different processes of breakdown and with different levels of toxicity. Second, understanding the movement of materials through the interstices of soil and within groundwater systems requires that we know a great deal about soil chemistries and composition, as well as about where groundwater is located, the size of aquifers, and the patterns of movement and recharge for groundwater systems. Finally, the nature of our society's use of materials and resources leads to enormous solid waste loads. Consequently, there is a lot of economic pressure to dump waste into landfills and other waste repositories. Since decisions in our society are shaped by a variety of perspectives, of which science is only one, our solutions to some problems (like storing solid waste) may lead to other problems (like contamination of soils and water from leaking landfills).

In this chapter, there are four research opportunities that allow you to study landfill problems.

- **Landfill Research Problem 1:** Groundwater Flow – Where Does It Go? 36
- **Landfill Research Problem 2:** Is the Water Safe to Drink? 47
- **Landfill Research Problem 3:** Pollutants, Plumes, and Permeability 61
- **Landfill Research Problem 4:** Likely Leachate Locations 72

Landfill Research Problem 1: Groundwater Flow – Where Does It Go?

Introduction and Advice

Before you start the research, examine **Study Notes 3.1** and **3.2** for some guidance about what you need to know before you start and what you should know when you are done. In **Study Note 3.3**, we provide a table you can use to keep track of assignments and their due dates.

> **Study Note 3.1: As you begin the research, you will need to have some prior knowledge in the following areas:**
>
> 1. A basic understanding of the scientific method and research design.
> 2. Knowledge about how landfills are constructed.
> 3. How and why to collect data on groundwater through groundwater monitoring wells.
> 4. Some knowledge of topographic maps and how they are constructed.
> 5. An understanding of factors that shape groundwater dispersion and retardation processes.

> **Study Note 3.2: What you should be able to do after completing the research.**
>
> 1. Enter a landfill, and collect groundwater quality and elevation data at each deep well and the leachate wells in the landfill.
> 2. Conduct a 3-dimensional analysis of topographic and groundwater surfaces in order to create a groundwater contour map.
> 3. Use the map to interpret patterns of groundwater flow and couple flow patterns to data on groundwater quality (pH, Specific Conductance, and Temperature) in order to predict impacts to down-gradient wells

> **Study Note 3.3: Use the table below to list the research assignment given by your instuctor. Make notes on this table about the types of work you will be expected to submit for a grade.**

Research Assignments	Specific Instructions for Completing Assignment	Completion Date
Phase 1		
Phase 2		
Other Assignments		

What's the Problem?

Take a simplistic approach. Dig a hole in the ground, partially fill it up with solid waste, throw some dirt on top to cover the whole mess, and call it a landfill. Now, imagine that it rains. Water tends to percolate down through the soil and, thus, into and possibly through the solid waste. What happens if the rainwater percolates through the solid waste, carries off soluble materials, and continues percolating down until it enters groundwater? Well, the simple story is that some of the materials carried from the landfill might enter the groundwater. The simple solution is to put a heavy duty liner in the hole before dumping the waste and spreading a waterproof cover over the landfill area.

Unfortunately, people didn't think about what happens as water moves through solid waste buried in soil, and many landfills used to be built without liners or covers. Besides, we can have problems without rain percolating through the landfill from the surface. Basically, water flows downhill. Water flows downhill on top of the earth, such as in streams and rivers, or even in rivulets from a rainstorm. Groundwater also flows downhill, but underground. Few places on earth are absolutely flat. Thus, we have the possibility of water moving laterally underground through a landfill and carrying off a variety of substances.

So, here's the problem. You want to know if groundwater is flowing through or under a landfill. Think about what you need to know in order to decide if groundwater is flowing though a landfill. First, you need some way to get to the groundwater, so we have to put in some sampling wells that are located in and around the landfill. Then think about where the groundwater might go. Remember that it will flow downhill. Finally, if in fact we find evidence that water is likely to move through the landfill, how will we determine if substances might be carried in the groundwater away from the landfill? One strategy for trying to solve this problem is to conduct research in two phases.

1. Collect groundwater quality and elevation data at each deep well and the leachate wells. Use these data to build a groundwater contour map.
2. Interpret groundwater flow and predict impacts to down-gradient wells. Check predictions with groundwater quality data (pH, Specific Conductance, and Temperature).

You can see these phases of research were chosen to answer our questions about where groundwater will flow and whether or not the groundwater carries substances leaking from the landfill. Read the Background Information and then head for the landfill. There's work to be done.

Background Information

The landfill you will be exploring has a long history of use, and it has no liner or leachate capture system. You will be using some of the environmental consultant's basic tools to decide whether this landfill is polluting groundwater. You will measure chemical and physical characteristics of the groundwater and of the polluting leachate. The techniques used are the cornerstone of almost all environmental investigations involving groundwater.

Want some advice? To conduct this research, you have to have some prior knowledge. **Figure 3.1** shows the current location of the landfill. In **Figure 3.2**, you can see the pre-existing valley that the landfill "filled up." You might want to review information on landfills from your text, and try to decide if the site of the landfill you're studying was a reasonable choice. Read about how landfills used to be constructed and how they are constructed now.

38   Chapter 3: Land Full of Landfills

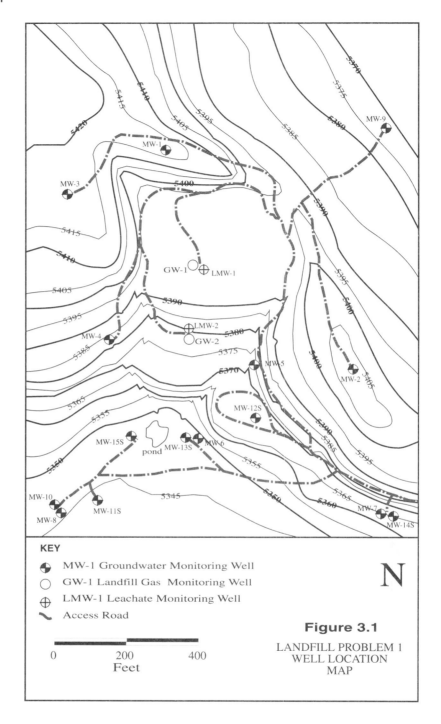

Figure 3.1
LANDFILL PROBLEM 1
WELL LOCATION
MAP

Certainly, you also have to know what groundwater is, and you will need to have some introductory knowledge of topographic maps. Before you start, make sure you are able to define elevation and understand how elevation is measured. It might be helpful for you to construct simple topographic maps by using elevation data and trying to use these data to draw the contours and landforms. Alternatively, get a topographic map of an area you know, and identify mounds and valleys while generally developing a picture in your mind of the terrain. It helps if you take a topographic map of an area near your institution, and look back and forth from map to the real world.

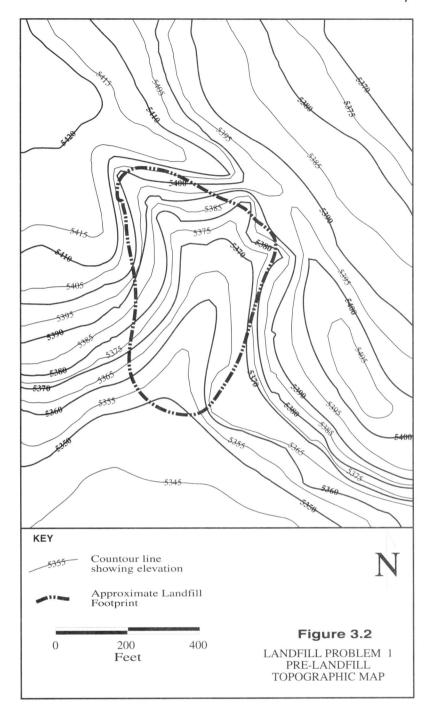

Figure 3.2
LANDFILL PROBLEM 1
PRE-LANDFILL
TOPOGRAPHIC MAP

In addition to some familiarity with landfill construction and topographic maps, make sure you know the definitions of the following:
- pH
- Alkalinity
- Specific Conductance

Chapter 3: Land Full of Landfills

These are important variables which we will be studying. If you don't know these terms, use your text to complete the following table.

Variable	Definition	Chemical Importance	What Factors Influence this Variable
pH			
Alkalinity			
Specific Conductance			

Finally, if you are entering the landfill as a research team of 2-4 students per computer, we recommend you assign tasks for each person. Have one person "drive" the mouse, one person navigate using the map, and one or two people individually record the group data on the sheet provided **(Table 3.1)**. One person can easily complete this lab, but you need to take a very organized approach. Regardless of how you complete this research problem, collect the data carefully and systematically, always checking to make sure your team members have written down the correct instrumentation readings.

Table 3.1: Site fluid variables. See variable definitions below the table.

Well	SUE (ft)	DTW (ft)	GWE (ft)	pH	SC (mho/cm2)	T (deg C)
MW-1	5418.23					
MW-2	5407.22					
MW-3	5417.89					
MW-4	5386.29					
MW-5	5387.56					
MW-6	5350.67					
MW-7	5368.25					
MW-8	5348.29					
MW-9	5373.65					
MW-10	5348.03					
LMW-1	5393.01					
LMW-2	5383.83					

Variable Symbol or Abbriviation	Description	Definition or Explanation
SUE	Stick-up elevation	Usually measured from top of well
DTW	Depth to water	Elevation of top of well that "sticks up" out of the ground
GWE	Groundwater elevation	SUE – DTW
pH	pH	Negative log of hydrogen ion concentration. Low pH's are acidic, high pH's are basic. Neutral fluidshave a pH of 7.
SC	Specific conductance (mho/cm)	Measures the ablilty of a fluid to conduct electricity
T	Temperature (degrees Celsius)	

Research Questions

You know the general problem. Is groundwater flowing through the landfill, and if it is, do we find any evidence that the groundwater is picking up substances from the landfill? Can you identify a specific research question that provides a focus for your efforts? Maybe there is more than one research question. For example, in **Phase 1**, you might ask, "How is groundwater flowing in and around the landfill?" In **Phase 2**, you could ask, "What are the likely impacts on wells at elevations lower than the landfill?"

You can see we took some real-world problems, looked at some background information, and now we try to formalize the problems in one or more specific research questions. Throughout the VRX simulations, you should focus on asking meaningful questions, and use these to create research designs that will answer the questions.

You will recall that you took a quick tour of the landfill simulation in the previous chapter. Refer back to that section if you need help getting started with **Phase 1** of this Research Problem

Phases Of Research

Phase 1 — Collect groundwater and leachate data in order to build a groundwater flow map.

Task 1:
Phase 1 is designed to answer the research question, "How is the groundwater flowing in and around the landfill?" You should use the map in **Figure 3.1** to help you find your way through the simulated landfill. If you are working in a team, assign one person to navigate and another to "drive." Begin your journey in the lower right corner of the map. Collect data at each deep well (MW-1 through MW-10) and at each of the two leachate wells (LMW-1 and LMW-2). The "S" or shallow wells are used in subsequent exercises — you need not collect any data at a well ending in "S."

At each well, collect the **depth to water**, which we abbreviate as DTW. Also, measure the pH, Specific Conductance (SC), and Temperature (T) of the fluid. Refer back to the volume titled **Instrumentation** in the *Bookshelf* for information about which instruments to use to collect the data and how to operate them. If you are working in a group, record the data in a form like the one provided in **Table 3.1**. The person recording the data should check each entry, with another team member also checking the transcription. If you are working in a team, remember to share the data among all research team members. Collect the well data in whatever order you want.

Task 2:
Now you need a tool that will help you answer the question, "How is groundwater flowing in and around the landfill?" With the data on DTW, you can calculate the groundwater elevation at each well, and then make an estimated groundwater elevation map. This kind of map is similar to a topographic map, except instead of showing the elevation of the land surface, it shows the elevation of the top of the groundwater underneath the land surface.

You will need to have **Table 3.1** and **Figure 3.3** handy. Consult **Research Note 3.1** for some advice about the steps to follow.

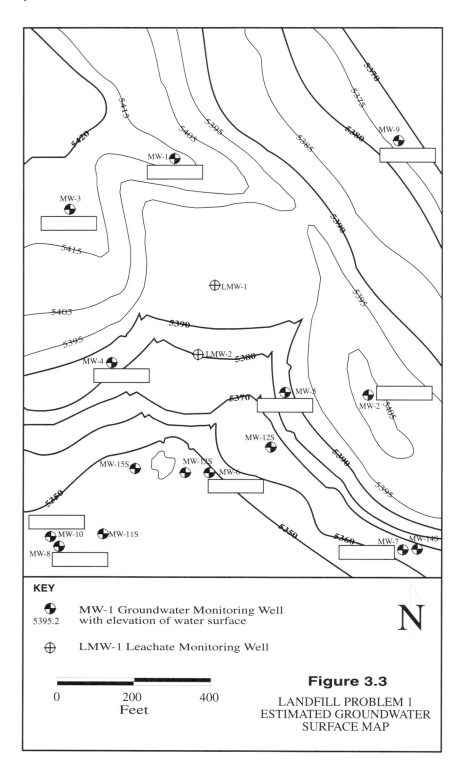

Figure 3.3
LANDFILL PROBLEM 1
ESTIMATED GROUNDWATER
SURFACE MAP

> **Research Note 3.1: Protocol for constructing a groundwater surface map.**
>
> 1. You measured the depth to groundwater from the top of the well, which is also known as the "stickup." To calculate the groundwater elevation at the well, subtract the measurement of depth to groundwater (DTW) from the stickup elevation (SUE) for all your measurements on **Table 3.1**. Round off your answer to the nearest tenth. (Example: 54.45 would round to 54.5, while 34.33 would round to 34.3)
>
> 2. Put these groundwater elevation numbers next to the corresponding wells on the map labeled **Figure 3.3**. Try to keep in mind that these numbers are the elevation of the top of the groundwater at each of the various wells.
>
> 3. Now you need some advice, because things are about to become difficult. You are going to draw lines on **Figure 3.3** that describe the groundwater surface. Two important facts that will help you are (1) the shape of the lines you will draw have the same shape as the topographic lines, and (2) these lines will connect points that have the same groundwater elevation. You should draw and label five lines that show the following elevations: 5320, 5330, 5340, 5350 and 5360.
>
> 4. Okay, now you have your five groundwater contour lines drawn in. You can use these lines as a tool to tell which way groundwater is likely to be flowing. Here's how it's done: about every half- inch or so on each groundwater line draw an arrow that is perpendicular to the line, with the arrow head in the direction towards lower elevations. These arrows point in the direction that a "chunk" of water would flow if it was sitting on the contour line where the arrow starts. Did you think there would be so many different directions ?

Phase 2 — Interpret how groundwater is likely to flow, and predict impacts on wells at elevations lower than the landfill.

Our central question for **Phase 2** was, "What are the likely impacts on wells at elevations lower than the landfill?" However, as we think about the landfill and review the background information, a number of related questions can be developed. We have organized some additional questions we hope will focus your efforts. Please answer these questions using the figures and tables you have constructed above. Also, take a step towards being a scientist, and write responses to the questions as if they were part of a technical report. In short, this means you should use complete sentences that capture and communicate the essence of your research findings.

44 Chapter 3: Land Full of Landfills

Task 1:

Look at **Figure 3.2**. This is the pre-existing valley that the landfill "filled up." The landfill was started long before EPA required liners, and raw waste was just piled on the ground in the valley. The valley floor defines the bottom of the waste pile. What would you say about the potential for groundwater pollution at this site if you thought the valley floor was bare, fractured rock?

_____
_____
_____
_____
_____
_____

Task 2:

You know the "shape" of the valley floor before the landfill was constructed (**Figure 3.2**), and you know what it looks like now (**Figure 3.1**). Draw some rough cross-sections of the waste pile.

W   E   N   S

Task 3:

The waste in each cell is generating leachate. The line below represents the surface that the waste cells were constructed on. Draw two layers of cells and show where leachate would pool in each cell.

North ＼
          ＼
            ＼_____ South

Task 4:

Look at the groundwater contour map you constructed (**Figure 3.3**). Following your arrows, in which directions does groundwater flow under the waste?

_____
_____
_____
_____
_____
_____

Task 5:
Compare the elevation numbers of the groundwater surface (**Figure 3.3**) that you constructed with the elevation numbers at the bottom of the waste pile (**Figure 3.2**). Is the waste in contact with groundwater anywhere? Back up your answer with numbers.

_____
_____
_____
_____
_____
_____

Task 6:
Examine the groundwater surface map you made (**Figure 3.3**). List two wells you think might be uncontaminated based on their location and direction of groundwater flow.

_____
_____
_____
_____

Task 7:
Study the groundwater surface map you made (**Figure 3.3**). List the wells that would intercept the groundwater that flows under the landfill.

_____
_____

Now look at your data (**Table 3.1**). Do the pH, Specific Conductance or Temperature readings of the groundwater in the wells you listed below suggest any leaks of leachate into groundwater? Use data to support your conclusion.

_____
_____
_____

Chapter 3: Land Full of Landfills

Task 8:

Groundwater sampling technicians normally sample wells at a landfill in sequence from the least-contaminated background wells to the most contaminated wells. List the order in which you would sample the wells listed below knowing what you now know.

_____ MW-9	_____ LMW-2	_____ MW-5
_____ MW-10	_____ MW-2	_____ MW-4

Task 9:

Let's examine one last question, which is the central question for **Phase 2** of the research. There is a town water supply well-field located one mile southeast of the landfill. Decide why you should or should not conduct additional studies in order to determine the risk of landfill contaminants reaching the town's water supply.

_____
_____
_____
_____
_____
_____
_____
_____
_____
_____

Congratulations! You have completed a typical exercise for which environmental consulting firms get paid **thousands** of dollars.

In The Know About Groundwater Flow

We posed two research questions, each associated with a phase of research. However, there were a number of related questions that evolved as we proceeded through the research. Imagine that you have to report your findings to a public interest group whose members are examining landfill issues in their county. This group is not a technical or scientific group, but ordinary American citizens concerned about the area in which they live. Answer the following questions, first in a technical format and then in a way that the voting public can understand. Do not oversimplify and create misconceptions about the research you conducted. Be particularly aware of the limitations of your research, especially in the context of making conclusions that will influence how people perceive landfill structure and function.

1. Why should we study the structure and function of solid waste landfills?

2. How do we access groundwater and collect data that provide information about the physical movement of groundwater and its chemical composition?

3. What are the reasons for building and interpreting a 3-dimensional analysis of topographic and groundwater surfaces?

4. How would we recognize if groundwater was contaminated and determine the direction of contaminant flow?

5. What factors lead to dispersion and retardation of groundwater flow?

Landfill Research Problem 2: Is the Water Safe to Drink?

Introduction and Advice

Before you start the research, examine **Study Notes 3.4** and **3.5** for some guidance about what you need to know before you start and what you should know when you are done. In **Study Note 3.6**, we provide a table you can use to keep track of assignments and their due dates.

> **Study Note 3.4: As you begin the research, you will need to have some prior knowledge in the following areas:**
>
> 1. How to access groundwater data through groundwater monitoring wells.
> 2. How to build and interpret a groundwater surface map.
> 3. A basic qualitative understanding of fluid mixing.
> 4. Some knowledge about groundwater dispersion and retardation processes.

> **Study Note 3.5: What you should be able to do after completing the research.**
>
> 1. Collect depth to groundwater data at selected wells and prepare a groundwater contour map.
> 2. Collect groundwater chemistry data with an understanding of environmental sampling procedures.
> 3. Interpret groundwater flow and water chemistry in order to predict impacts on wells at lower elevations that are within any plumes of leachate flow.
> 4. Prepare a technical report that offers a regulatory argument with supporting data.

What's The Problem?

Let's keep working with the landfill we studied in **Research Problem 1**. By the time we had completed the research related to **Problem 1**, we were able to make some conclusions about groundwater flow and the possible contamination of groundwater by leachate from the landfill. However, a striking feature of the real world is that it is spatially and temporally heterogeneous. This simply means the real world has lots of different kinds of habitats and environments. Furthermore, any given location in the real world tends to change through time. So, there is variation in space and variation in time.

Part of the complexity of conducting research is due to the difficulty and expense of studying the same place for long periods of time. As you might imagine, it is even more difficult to study many different places simultaneously for extended periods of time. However, let's focus on the landfill we studied in Problem 1. Do we know everything there is to know about this landfill? Frankly, we have hardly begun to get at its complexity. Imagine that the last several months have been wetter or drier than usual. There may be changes in the groundwater level. Furthermore, suppose there are toxic materials in the landfill, and some of these materials have been moving slowly through the soil and into the groundwater. Such toxic substances may not have been detectable a month ago, but are now present in the groundwater and may pose a threat to people drinking from wells supplied by the groundwater.

Study Note 3.6:

Use the table provided below to list the research assignments given by your instructor. Make notes on this table about the types of work you will be expected to submit for a grade.

Research Assignments	Specific Instructions for Completing Assignment	Completion Date
Phase 1		
Phase 2		
Phase 3		
Other Assignments		

Chapter 3: Land Full of Landfills

You can check the literature, and you will find that benzene and arsenic are problems in many landfills, particularly those that have received industrial and construction wastes. These may leach into groundwater and flow as plumes of contaminants in groundwater. As an experienced researcher, you are assigned to revisit the landfill and continue studies on the groundwater and what the groundwater contains. There may be arsenic and benzene in the groundwater, moving with the water through the aquifer. The research could evolve through three phases.

1. Collect depth to groundwater data at selected wells, and prepare a groundwater contour map.
2. Collect groundwater chemistry data, focusing on arsenic and benzene.
3. Interpret groundwater flow and water chemistry in order to predict impacts on wells at lower elevations that are within any plumes of leachate flow.

Background Information

You have been brought in as a special consultant to answer some questions the State Department of Environmental Quality has about this landfill. A neighbor of the landfill, who lives 2,000 feet to the east, has complained that his water well is being contaminated by the landfill. The DEQ sampled his well and found that the groundwater contained benzene, which is a known carcinogen, and arsenic, a toxic metal. Both of these chemicals are regulated by the EPA, and both have been found at the landfill in the past. As part of your contract, DEQ requires you to write a short report indicating whether the landfill could be contaminating the neighbor's well. The DEQ memo outlining the previous work can be found as **Attachment 3.1** at the end of this laboratory section.

This research project requires an introductory knowledge of groundwater contour maps and basic groundwater chemistry data. In regards to chemistry, the researchers should have a basic understanding of pH, Specific Conductance, Temperature, and why arsenic and benzene are toxic. Knowledge of units is also crucial. For example, the researchers need to have a working knowledge of fluid concentration units such as milligrams per liter and parts per million. Although not absolutely essential, the research team members will find a knowledge of geometry useful. Finally, in order to conduct the research for **Problem 2**, the research team members should have experience contouring groundwater elevation data, constructing simple groundwater contour maps, and interpreting groundwater flow directions from contours.

Remember, you have to write a report for DEQ. This report should have a technical format and succinctly describe your research findings. Check with your instructor about the exact format for this technical paper.

Research Questions

The basic problem is similar to that studied as **Problem 1**. We need to know if groundwater is flowing near, through, or under the landfill. If it is, do we find any evidence the groundwater is picking up substances from the landfill? In **Problem 2**, we are specifically concerned with toxic substances that may contaminate wells supplied by the same groundwater system. Again, can you identify specific research questions that provide a focus for your efforts? For example, in **Phase 1**, you might ask, "How is groundwater flowing in and around the landfill?" In **Phase 2**, you could ask, "What are the chemical characteristics of the groundwater at different locations in and around the landfill?" Finally, for **Phase 3**, you might ask, "Is there sufficient evidence to conclude that the landfill could be contaminating the neighbors well?"

As in **Problem 1**, we took some real-world problems, looked at some background information, and are now trying to formalize the problems into three specific research questions. You might also note that in **Problem 2**, the data for **Phase 3** is provided by the research in **Phases 1** and **2**. However, each of the three phases of research must be completed and discussed in order to write a substantive technical paper for DEQ. In fact, you might find it useful to treat the results of your research in each of the phases as a separate section in the technical paper.

Phases Of Research

Phase 1 — Collect data at the shallow wells and at leachate well LMW-2, and prepare a groundwater contour map.

Task 1:
Remember that **Phase 1** is designed to ask the question, "How is groundwater flowing in and around the landfill?" Once again, you will be constructing a groundwater surface contour map. So, collect depth to groundwater (DTW) data at wells MW-11S through MW-15S and at leachate well LMW-2. Record the data on **Table 3.1**, and complete all of the calculations. Calculate the amount of water you need to withdraw from the well to ensure you are sampling fresh aquifer water. This is done by calculating a well volume, which is the amount of water in the well casing. Copy the well volume you calculated for each well in **Table 3.2** into the appropriate location in **Table 3.3**.

Task 2:
Calculate the groundwater elevation at each shallow well, and then make a groundwater contour map for the shallow aquifer. You will need the groundwater flow map to support your conclusions. In case you feel a little tentative with groundwater maps, the general directions are shown again in **Research Note 3.2**.

Research Note 3.2: Constructing a groundwater surface contour map.

1. Subtract the measurement of depth to groundwater (DTW) from the stickup elevation (SUE).
2. Put these groundwater elevation estimates next to the corresponding wells on your map labeled **Figure 3.4**.
3. Draw and label the lines of equal groundwater elevation on the map.
4. Draw the arrows that show the probable direction of horizontal groundwater flow. Look at your map, and decide if you can predict which wells will show the greatest impacts from the landfill.

Chapter 3: Land Full of Landfills

Phase 2 — Collect groundwater chemistry data.

In **Phase 2**, we are interested in the question, "What are the chemical characteristics of the groundwater at different locations in and around the landfill?" You will be using a pump device to sample groundwater. If you are working in a group, make sure everyone has completed the calculations for well volume and has copied them to **Table 3.3**. You will record data at each of the three first whole well volumes, and then again when stabilization has occurred. Be sure to sample the wells in this order: MW-11S, MW-15S, MW-14S, MW-12S, MW-13S, and finally the leachate well LMW-2. **Research Note 3.3** provides some guidance for sampling groundwater.

Research Note 3.3: Groundwater sampling instructions.

1. Set-up the pump, and start purging groundwater from the well. Measure and record the pH, Specific Conductance, and Temperature at each whole well volume. DEQ requires you to show these data in your report. When these parameters have stabilized, you are ready to sample. You will know when stabilization has occurred if you watch your instrument readings carefully. Remember to record the final readings.

2. Once you have confirmed that the water has stabilized, collect a groundwater sample by turning the valve marked discharge/sample. If you take a sample before the groundwater chemistry has stabilized, your results may not be truly representative of the aquifer water chemistry, and you might come to the wrong conclusions.

When you leave the landfill, your samples will be automatically sent to an EPA approved laboratory. The results will appear on the closing screen, along with the costs for your research effort and the laboratory analyses. Write these analyses down, because they will disappear after you leave the program.

Phase 3 — Synthesize groundwater flow and chemistry data, and prepare a report for DEQ.

Task 1:

For **Phase 3**, we are focusing on the question, "Is there sufficient evidence to conclude the landfill could be contaminating the neighbor's well?" As mentioned, the chemistry information is presented by the program when you finish sampling and quit the program. Record your groundwater chemistry results, and the cost of the work in the spaces provided on **Table 3.3**. If you sampled a well more than once, then record the data you feel is correct.

Task 2:

If you are working in a group, this would be a good time to discuss the facts of the case before everybody gets away. **Research Note 3.4** provides questions that serve as framework for your synthesis.

Research Note 3.4: Questions that serve as a framework for developing your report to DEQ.

1. What is the direction(s) of shallow groundwater flow at the site?

2. What are the concentrations of arsenic and benzene you found in the shallow aquifer and in the landfill leachate?

3. When you compare leachate wells to groundwater wells, do you find huge differences in the concentrations of contaminants? What does this say (if anything) about how much leachate is mixing into groundwater?

4. What differences in concentration of benzene and arsenic do you see as you examine the wells closest to the landfill and those farthest away? (Consult your flow map!) Are there any trends in concentration?

5. Is the neighbors well in the same aquifer as the shallow aquifer at the landfill ? Compare **Figure 3.5** with the information given in **Attachment 3.1** when answering this question."

6. How do the landfill groundwater well concentrations of benzene and arsenic compare to the concentrations found in the neighbor's well?

7. Is it likely the landfill is responsible for polluting the neighbor's well? Which of your findings support your conclusion?

Table 3.2: Site fluid variables. See variable definitions below the table.

Well:	SUE (ft)	DTW (ft)	GWE (ft)	WBE (ft)	L (ft)	WELL VOLUME (g)
MW-11S	5347.22			5330.99		
MW-12S	5371.25			5333.03		
MW-13S	5350.65			5332.13		
MW-14S	5368.01			5329.40		
MW-15S	5351.72			5330.10		
LMW-2	5383.83			5356.07		

Variable Symbol or Abbreviation	Description:	Definition or Explanation
SUE	Stick-up elevation (feet).	This refers to the elevation of the top of the well which "sticks up" out of the ground.
DTW	Depth to water (feet).	This is usually measured from the top of the well.
GWE	Groundwater elev. (feet)	= (SUE - DTW) Well
WBE	Well bottom elevation.	The lowest part of the well.
L	Height of water in a well (feet)	
Well Volume	The volume of fluid in a well	L x 0.163 gallons/ft. (Assumes a well pipe diameter of 2 inches.)

Table 3.3: Table for collecting well data.

Well:	Purged Volume (gal)	pH	SC (μmho/cm)	T (deg C)	Benzene mg/L	Arsenic mg/L
MW-11S	WV * 1 =					
WV =	WV * 2 =					
	WV * 3 =					
	FINAL =					
MW-15S	WV * 1 =					
WV =	WV * 2 =					
	WV * 3 =					
	FINAL =					
MW-14S	WV * 1 =					
WV =	WV * 2 =					
	WV * 3 =					
	FINAL =					
MW-12S	WV * 1 =					
WV =	WV * 2 =					
	WV * 3 =					
	FINAL =					
MW-13S	WV * 1 =					
WV =	WV * 2 =					
	WV * 3 =					
	FINAL =					
LMW-2	WV * 1 =					
WV =	WV * 2 =					
	WV * 3 =					
	FINAL =					
					Total Cost	

Attachment 3.1: Memorandum from the Department of Environmental Quality.

Department of Environmental Quality
Groundwater Division
INTERNAL MEMO: January 12, 1998

RE: Results of sampling at Gunn Residence

On January 3, 1998, Department of Environmental Quality (DEQ) staff sampled the drinking water supply well of Mr. Robert Gunn, 1020 West Landfill Drive, in response to a series of complaints by Mr. Gunn. According to DEQ records, Mr. Gunn first contacted DEQ's local office in June 1997, then again in November 1997, to complain about a change in the water quality at his residence. Mr. Gunn operates Gunn's Auto Repair adjacent to the residence.

Mr. Gunn has indicated that it is his belief that the Solid Waste Landfill, located approximately 2000 feet to the west of his residence has impacted the groundwater around his well. DEQ hydrogeologists have reviewed groundwater flow maps for the region between the landfill and the Gunn residence, and have determined that groundwater in some places flows east from the landfill towards the Gunn residence. DEQ staff reviewed the well-driller's log for Mr. Gunn's well, which is shown below:

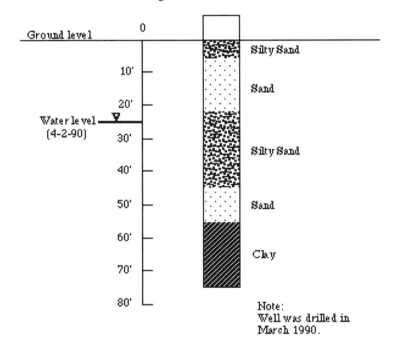

GUNN Residence Well Log

Based on the analysis of the well log, it appears that the Gunn residence well is completed in the shallow alluvium, and does not penetrate the bedrock aquifer. DEQ staff examined the well, which is located in a protected enclosure outside the Gunn residence. No visible signs of contamination around the wellhead were observed, and the well access hole was sealed with a stainless-steel bolt. The pump delivery system appeared to be clean and functional. Reportedly the well contains a 3.5 inch Grunfoss submersible pump, which delivers approximately 10 gallons per minute for sustained rates of up to 12 hours.

A depth to water measurement of 25 feet (bgs) was taken using a ACTAT well probe meter. Since the well had reportedly been used in the last 24 hours, this measurement likely reflects a water table that is recovering from pumping. The total depth of the well was also measured at 75 feet.

DEQ staff attached a 10 foot piece of new, clean Tygon plastic tubing to the well cut-off valve at the well head and proceeded to purge the well at approximately 5 gallons per minute. Approximately 300 gallons were purged prior to sampling. Samples were collected for total metals and volatile organic compounds. The samples were refrigerated and delivered to DEQ laboratories at the regional office at 4:00 p.m. the same day.

The analytical data received from the lab are shown below in **Table 3.2**.

Table 3.2. Gunn Sampling Results

Analyte	Value
pH	7.3
SC	320 μmho/cm
Temp (C)	19.1
Arsenic	0.02 ppm
Calcium	49 ppm
Magnesium	25 ppm
Acetone	0.4 ppm
Benzene	0.1 ppm
Ethylbenzene	0.3 ppm
Toluene	1.1 ppm

It is the recommendation of DEQ staff that a professional environmental consultant be hired to sample the shallow groundwater wells completed in the alluvium and the lower leachate well at the Solid Waste Landfill. The samples should be analyzed for Benzene and Arsenic, which are the two contaminants of concern in Gunn's well. The consultant should determine the direction of shallow groundwater flow at the landfill, and present DEQ with a report indicating whether it is likely, based on the evidence found, that the Solid Waste Landfill is contaminating the Gunn residence groundwater well. The consultant's fee should be limited to $1,200.00 for fieldwork and sample analyses.

In The Know About Where Contaminants Go

We posed three research questions, each associated with a phase of research. You have completed the research and filed a report with DEQ. They call and arrange for you to present your findings to a panel of technical experts. This panel asks you the following questions in order to see if you can defend your research results. Respond to the questions in a technical format, providing data to support your conclusions.

1. Why do real-world systems vary through time and space? Are landfills susceptible to such changes?

2. What data are crucial for building and interpreting a groundwater surface contour map?

3. How do we couple a 3-dimensional analysis of groundwater surface topography and groundwater chemistry to understand movement of substances from a landfill to neighboring areas?

4. How do you craft an argument for a regulatory agency that provides substantive data, reasonably analyzed, interpreted, and developed into coherent conclusions that can be used to make decisions about environmental systems?

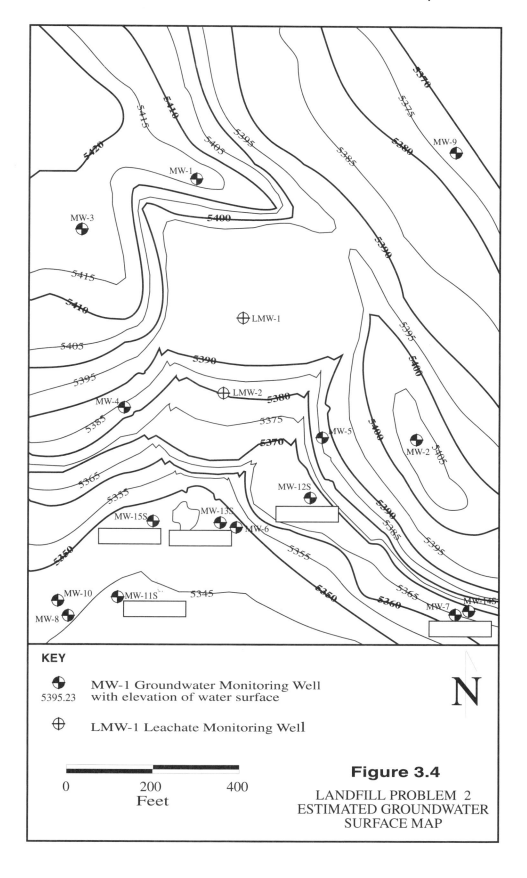

Figure 3.4
LANDFILL PROBLEM 2
ESTIMATED GROUNDWATER
SURFACE MAP

Chapter 3: Land Full of Landfills

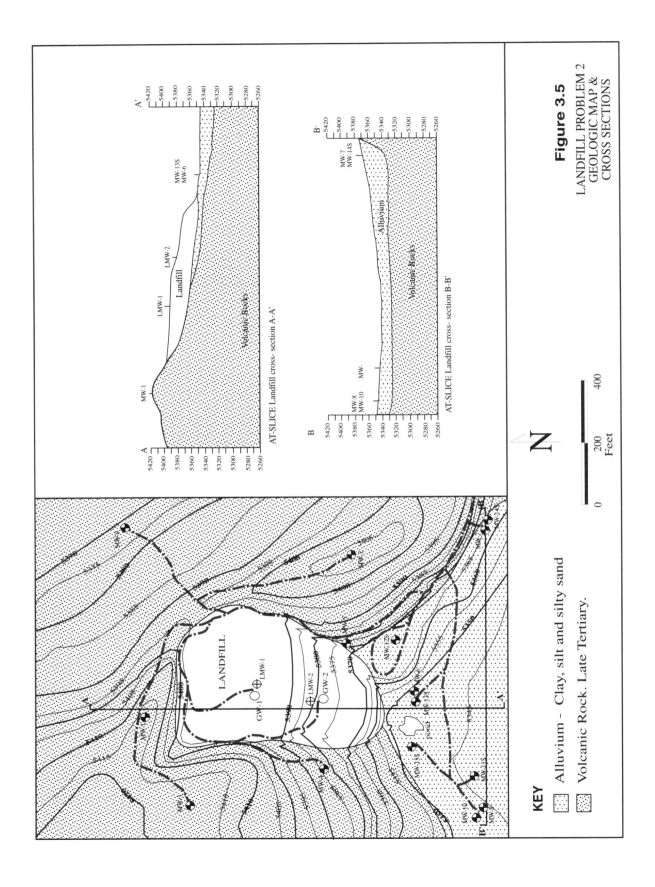

Figure 3.5
LANDFILL PROBLEM 2
GEOLOGIC MAP &
CROSS SECTIONS

Landfill Research Problem 3: Pollutants, Plumes, and Permeability

Introduction and Advice

Before you start the research, examine **Study Notes 3.7** and **3.8** for some guidance about what you need to know before you start and what you should know when you are done. In **Study Note 3.9**, we provide a table you can use to keep track of assignments and their due dates.

Study Note 3.7: As you begin the research, you will need to have some prior knowledge in the following areas:

1. How to access groundwater data near landfills through groundwater monitoring wells.
2. How to build and interpret a groundwater surface map.
3. A basic qualitative understanding of fluid mixing.
4. Some knowledge about groundwater dispersion and retardation processes.
5. Some knowledge about groundwater adsorption and precipitation processes.

Study Note 3.8: What you should be able to do after completing the research.

1. Collect depth to groundwater data at selected wells in two aquifers, and prepare a groundwater contour map.
2. Design a sampling plan to collect groundwater chemistry data.
3. Interpret groundwater flow and water chemistry in order to predict impacts on wells at lower elevations that are within any plumes of leachate flow.
4. Prepare a technical report on your findings.

> **Study Note 3.9:** Use the table provided below to list the research assignments given by your instructor. Make notes on this table about the types of work you will be expected to submit for a grade.

Research Assignments	Specific Instructions for Completing Assignment	Completion Date
Phase 1		
Phase 2		
Phase 3		
Other Assignments		

What's The Problem?

You remember from our introduction to **Research Problem 2** that we talked about how any given location in the real world tends to change through space and time. We called this variation spatial and temporal heterogeneity. In other words, there is variation in space and variation in time. By studying groundwater flow and concentrations of chemicals in the groundwater, we were able to make some conclusions about the likelihood of leachate moving into the groundwater and traveling with the groundwater to sites distant from the landfill. The design of our study did not include repeating our measurements over a long period of time. Nor did we look at any other variables besides the elevation gradient and a few chemical constituents.

Let's introduce another level of complexity. We talked about the possibility of leachate entering groundwater and flowing as plumes of contaminants in the groundwater. We also pointed out that there may be changes in the groundwater contour because of variations in precipitation. However, what about the soil and other geological units through which the groundwater flows? Suppose these are pretty stable through time (at least on the order of hundreds of years) but vary through space? A simple example would be that the landfill was constructed in an area that had several different types of geological structures. So, as groundwater moves through the landfill and surrounding areas, it runs into and then through or around different soils, rocks, and other geological formations.

Some of these formations may alter the flow rates and also the chemical composition of the groundwater. For example, water will flow faster through soils that are more porous, but cannot flow at all through certain types of bedrock. In regards to chemistry, groundwater may dissolve substances from surrounding soils and rock materials, but can also lose substances through a variety of physical processes, like filtering, or chemical processes, such as precipitation and sequestering.

Think about all this for a moment. Even if the groundwater levels stayed the same, there would be spatial variation in the geological formations the groundwater encountered as it flowed. Now, suppose the groundwater levels changed through time, as often happens seasonally because of changes in precipitation. To get your brain going, also imagine there are changes in the quantity and kinds of solid waste placed in the landfill. Put all of these changes together, and you have possible variation in the leachate quality and quantity, variations in the groundwater level and flow rates, and variation in soils and rocks through which the groundwater is flowing. You, the scientist, have to study this system, and make conclusions which may very well influence the quality of human life in neighborhoods that can be a considerable distance from the landfill site. To overstate this point a bit, you may be betting someone's life on your conclusions.

One interesting, and somewhat intricate, research problem is to determine if there are contaminants from a landfill in the groundwater flowing near the landfill site, and, if there are contaminants, to define the shape of the plume of these contaminants. To get an idea of what we mean by plume, think about real-world vistas or pictures you have seen of a muddy stream entering a clear lake or the ocean (like the Mississippi entering the Gulf of Mexico). Often, there is a visible plume formed by the muddy stream as it mixes with the other body of water.

Carrying this analogy into our groundwater system, suppose that leachate was seeping through the soils and into the groundwater. As the leachate mixed and began to flow with the groundwater, it would form a plume. In a simple sense, a contaminant plume is a three dimensional volume within the groundwater that contains contaminants. As in the case of a muddy stream entering a clear lake, there is a dilution effect, which is called dispersion. By dispersion, we mean the contaminant is dispersed into a larger volume of water, and is therefore diluted.

In the case of a stream entering a lake, the water turbulence in the larger body of water will often disperse the materials carried into it by the stream and the plume will disappear. There is also mixing in groundwater systems. Contaminants can be dispersed throughout the whole volume of the groundwater. Often, however, the contaminant plume is a three dimensional shape within the larger volume of the groundwater.

We will provide a scenario in which groundwater may be contaminated by two chemicals. You will have to find out where the contamination is located, and why the contamination is limited to some areas but not others. As in **Problems 1** and **2**, we suggest you outline the phases of research. The central problem is that a landfill appears to have leachates that are entering the groundwater, and we need to know the extent of the contamination. Since the groundwater may be flowing down an elevation gradient, one phase should focus on developing a groundwater contour map. Since we have specific contaminants we will be studying, a second phase should sample groundwater for the presence of these chemicals. Finally, we will need to synthesize our findings into a coherent story, and this will constitute a third phase of work. In terms of specific tasks, the phases can be defined as follows.

1. Collect depth to groundwater data at selected wells, and prepare a groundwater contour map.
2. Collect groundwater chemistry data.
3. Interpret groundwater flow and water chemistry in order to predict impacts on wells at lower elevations that are within any plumes of leachate flow.

We will let these phases stand for now, and return to them after we look at some background information and define specific research questions.

Background Information

The research team is assigned to study the contaminant plumes at a landfill. The landfill sits atop several different geologic units, and each unit has different properties that affect how the contaminant can pass through it. There are several processes that decrease contaminants in groundwater as it flows through an aquifer. Remember, the process of dispersion tends to dilute the concentration of a contaminant as it flows from the source (the landfill) to the measuring point (a well). Some organic contaminants tend to be sequestered by carbon-rich rock layers through the process of adsorption. In other cases, the pH of groundwater can affect whether a metal contaminant is soluble or undergoes the process of precipitation.

Take a look at **Figure 3.8** which shows the geologic map and cross-sections. Three different geologic units occur in the landfill vicinity. The two deep units that make up the bedrock aquifer are both volcanic rocks. Although there are two geologic units that make up the bedrock, they behave together like one aquifer. **Research Note 3.5** provides an overview of the geological units.

> **Research Note 3.5: Overview of geological units in Research Problem 3.**
>
> The **basalt** is primarily a lava flow, with low permeability but many cooling fractures along which groundwater can move. Water in contact with this rock type is generally neutral in pH. Overall this rock has a moderate permeability.
>
> The **silicic debris flow deposits** are made of glassy volcanic particles, mostly sand and silt size. Water in contact with this rock type becomes strongly alkaline. Permeabilities are also moderate in this rock.
>
> The third geologic unit is the **alluvium**, which by itself forms the shallowest aquifer (see the cross-sections). This alluvium is a layered sequence of sands, clays, and richly organic layers called "peat." Groundwater in this aquifer tends to be a bit more acidic. The alluvium has a much higher permeability than the bedrock.

Your job in this laboratory will be to figure out where the contaminant plumes are, and why they are in those locations. The landfill was used to dump a large number of circuit boards and the waste from their manufacture. This waste includes both trichloroethylene (TCE), an organic solvent, and copper (Cu). We will investigate the passage of these two contaminants through the aquifers near the landfill. To do this, you will test each well to determine the depth to water.

With this data you will construct a groundwater surface map in order to determine the direction of groundwater flow. Next you will decide on which wells to sample for TCE and Cu concentrations, remembering that sampling for this purpose takes time and money. This sampling requires that you pump water from the wells for a sufficient period so the water you actually collect is representative of what is in the groundwater, rather than simply what is in the well itself. TCEs, for instance, can migrate out of the water into the air, so your initial readings for that compound might be low if you only sampled water at the well's surface.

Choose enough of the wells to sample so you can determine how the plumes behave in each geologic unit, but do not sample every well (unless you have lots of time on your hands and a pile of money). Remember if you were doing this as a consultant, you would have to keep to a contracted budget amount.

Using this information, you will then produce a short technical report that describes the contaminant plumes, and why the contaminants behave as they do in these aquifers.

In regards to prior knowledge, we are trusting that you have an intermediate knowledge of groundwater contour maps from your work on **Problems 1 and 2**. Likewise, you should be familiar with aquifer structure and have a basic understanding or groundwater chemistry. For **Problem 3**, the research team will also have to work with geologic maps, and will need a basic understanding of volcanic rocks as well as the basic types of sedimentary rocks. You might want to review the chemical concepts of adsorption and precipitation as well as the units used for chemical concentration. In terms of problem solving skills, the research team will have to think both quantitatively and qualitatively in order to recognize negative and positive trends.

Research Questions

It is time to focus our approach by refining our understanding of the problem and posing specific questions we can answer by conducting research tasks. We are going to take a slightly different approach than we used in **Problems 1** and **2**. Remember, the question drives the research. The overall question is, "Are there contaminants in the leachate that are entering groundwater and posing a threat?" We said above that, first, we need to know something about groundwater flow, basically answering the question, "How does groundwater flow in and around the area of the landfill?" So, **Phase 1** should focus on creating a groundwater contour map. A point of interest and concern, however, is there are different geological units in the landfill area. Their presence does not change the question, but we do have to think carefully about which wells to sample in order to make sure we cover all possible paths for groundwater flow.

Phase 2 was designed to answer the question, "What is in the groundwater in the different areas within and around the landfill?" Remember that we have to pump out each well until we are sure we are sampling groundwater, but which wells should we sample? We have specific needs for data on concentrations of TCE and Cu, and we have to watch our budget. You will have to think carefully about how many wells to sample and which wells to sample. Then, you will have to collect the important water chemistry data at each of these chosen sites.

Finally, **Phase 3** is devoted to interpreting the results of **Phases 1** and **2**. This process of data analysis and interpretation will allow us to answer the overarching questions, "Are there contaminants in the leachate that are entering groundwater and posing a threat?"

Phases Of Research

Phase 1 — Collect data and build the groundwater surface contour maps.

Task 1:
Remember that the overarching research question we posed was, "Are there contaminants in the leachate that are entering groundwater and posing a threat?" In **Phase 1**, we will be focusing on, "How does groundwater flow in and around the area of the landfill?" This means we will once again be constructing a groundwater surface contour map.

Collect DTW data at each of the groundwater wells and at the two leachate wells. Record the data on **Table 3.4**, and complete the calculations for groundwater elevation. Using the groundwater elevations, draw groundwater surface maps for the bedrock and alluvial aquifers (**Figures 3.6** and **3.7**), and draw small arrows on each contour line (every inch or so) to indicate directions of water movement. The arrows point in the down-gradient direction and are perpendicular to the contour line. Return to **Research Note 3.1** or **3.2** to refresh your memory about how to construct a groundwater surface contour map.

Phase 2 — Devise a sampling plan and collect the groundwater chemistry data.

Task 1:
In **Phase 2**, we turn to the question, "What is in the groundwater in the different areas within and around the landfill?" Now comes the fun (deep thinking) part. You know the groundwater flow directions in the two aquifers, and you have the geologic map and cross-sections of the area to show you where each geologic unit is located. Decide which wells you should sample in order to see how the concentrations of the two contaminants (TCE and Cu) vary in the three geologic units. Keep in mind that all this costs money, so although you could sample each well, there may not be enough money to pay for the work. Place a check mark next to the wells on **Table 3.4** that you plan on sampling.

Task 2:
In order to collect water samples, you have to get rid of water that has been sitting in the well. That is, you have to pump out water until you get to the "fresh" groundwater. You will use the pump to sample the groundwater. Research Note 3.6 provides some guidance for this procedure.

Task 3:
Your water samples will be submitted to the laboratory, and the analyses will be made available to you before you quit the program after you are done sampling. Write these analyses down — they will disappear after you leave the program.

Research Note 3.6: Procedure for making sure you sample groundwater and not water that has been standing in the well pipe.

1. Select the pump, and start purging groundwater from the well. Watch the pH, Specific Conductance, and Temperature carefully as you pump out water. When these parameters have stabilized, you are ready to sample. Remember that time is money, so you should not just let the pump run forever. On the other hand, if you take a sample before the groundwater chemistry has stabilized, your results may be not truly representative of the aquifer water chemistry, and you might come to the wrong conclusions. Record the final readings including how much water you purged.

2. Collect a groundwater sample by turning the valve marked discharge/sample. Water fills a sampling bottle and will automatically be sent to the laboratory for analysis. Then move on to the next well, repeating the above process until you have sampled all of the wells in your sampling plan.

Phase 3 — **Synthesize your results into a coherent report that discusses the likely contamination of groundwater, the movement of contaminants, and the potential impacts on neighboring groundwater systems.**

Task 1:
You need to write a coherent results section that contains data summaries and analyses. For example, the two groundwater contour maps you made and data summaries of **Table 3.4** will provide crucial support for your discussion about the presence of contaminants and behavior of the contaminants in the groundwater.

Task 2:
The report needs to address whether or not the two contaminants behave differently as they travel through the three geologic units. Read your text or research the literature for information on how contaminants behave in different environments and at various pHs. Can such information allow you to describe the probable reasons for the contaminant behavior?

Task 3:
The final set of discussions in your report should address the likely processes involved in creating the contaminant distribution you discovered. Include a short explanation of the likely chemical reactions involved and the aquifer characteristics. Outline how these characteristics control the groundwater movement. This process of synthesizing an understanding from research data will require some thought. We encourage you to treat your classmates as colleagues. Also, do not be shy about discussing the research with your instructor. In the end, however, remember that any work you submit for grading must be your own or must conform to instructions from the faculty member teaching your course.

In The Know About Pollutants, Plumes, And Permeability

Imagine that your work is of high quality and you would like to publish your results in a scientific or technical journal. As a starting point, go to the library or an on-line service, and search for journal articles related to your research. Study the format of the papers printed in these journals. Second, write short technical pieces that answer the following questions. Compare your writings to journal articles on similar subjects.

1. Make a list of different soils and rocks in the landfill. How is each likely to influence groundwater adsorption and precipitation processes for TCE and Cu?

2. Switch the positions of the different geological units. How will this effect the distribution of the TCE and Cu?

3. Examine your research design. If you had twice as much money and time was not a factor, how would you change your design?

4. Is there any other way to study the TCE and Cu concentrations, distribution, and movement without a groundwater contour map?

Table 3.4 Landfill Groundwater Data

Well:	SUE (ft)	DTW (ft)	GWE (ft)	pH	SC (µmho/cm²)	Temp (C)	Purge (gal)	Cu (mg/L)	TCE (mg/L)
MW-1	5418.23								
MW-2	5407.22								
MW-3	5417.89								
MW-4	5386.29								
MW-5	5387.56								
MW-6	5350.67								
MW-7	5368.25								
MW-8	5348.29								
MW-9	5373.65								
MW-10	5348.03								
MW-11s	5347.22								
MW-12s	5371.25								
MW-13s	5350.65								
MW-14s	5368.01								
MW-15s	5351.72								
LMW-1	5393.01								
LMW-2	5383.83								

SUE = Stick-up elevation, the elevation of the top of the well
DTW = The depth to water measured by the meter from the top of the well
GWE = Groundwater elevation = (SUE - DTW)
Purge = The volume of water in gallons pumped from the well before taking a sample

70   Chapter 3: Land Full of Landfills

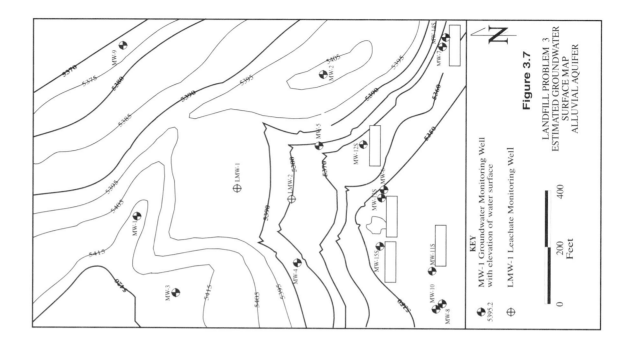

Figure 3.7
LANDFILL PROBLEM 3
ESTIMATED GROUNDWATER
SURFACE MAP
ALLUVIAL AQUIFER

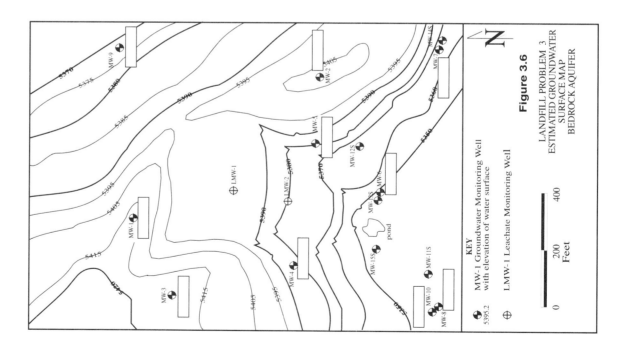

Figure 3.6
LANDFILL PROBLEM 3
ESTIMATED GROUNDWATER
SURFACE MAP
BEDROCK AQUIFER

Virtual Reality Excursions 71

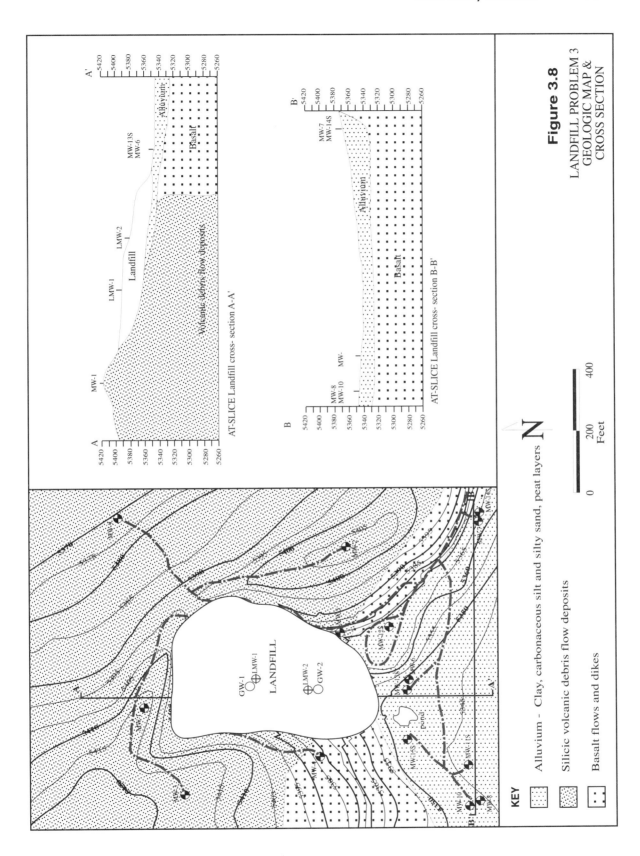

Figure 3.8
LANDFILL PROBLEM 3
GEOLOGIC MAP &
CROSS SECTION

Landfill Research Problem 4: Landfill Gas Migration

Introduction and Advice

Before you start the research, examine **Study Notes 3.10** and **3.11** for some guidance about what you need to know before you start and what you should know when you are done. In **Study Note 3.12**, we provide a table you can use to keep track of assignments and their due dates.

> **Study Note 3.10: As you begin the research, you will need to have some prior knowledge in the following areas:**
>
> 1. How to access groundwater and gas data near landfills through groundwater monitoring wells.
> 2. How to build and interpret a groundwater surface map.
> 3. A basic qualitative understanding of gas flow through soils and rocks.
> 4. Some knowledge about processes that facilitate or retard gas migration.

> **Study Note 3.11: What you should be able to do after completing the research.**
>
> 1. Develop testable hypotheses for patterns of gas migration in a landfill.
> 2. Collect gas and groundwater data from the landfill.
> 3. Build graphic tools for studying relationships between gas concentrations and distance from suspected sources.
> 4. Synthesize data and evaluate hypotheses.

Study Note 3.12: Use the table provided below to list the research assignments given by your instructor. Make notes on this table about the types of work you will be expected to submit for a grade.

Research Assignments	Specific Instructions for Completing Assignment	Completion Date
Phase 1		
Phase 2		
Phase 3		
Other Assignments		

What's The Problem?

In the three previous Research Problems, we examined leachates. There are other substances that can cause problems in landfills. For example, decomposition of organic materials can lead to production of methane and hydrogen sulfide. Carbon dioxide and carbon monoxide can also be produced by a variety of chemical and biological reactions within a landfill. These gases can permeate the landfill and build up to toxic or explosive levels.

In fact, one of the authors of this book was sampling a monitoring well in a landfill and found a dead bird in the well. While wondering how the bird got there, he noticed another bird land on the lip of a monitoring well and immediately topple over into the well. This second bird had been overcome by the gases in only a few seconds, almost certainly meeting the same fate as the first bird. While you might not succumb as easily as those birds, high levels of toxic gases should be approached very cautiously and only with appropriate protective gear.

In this research problem, we will explore the distributions, concentrations, and migration of gases in a landfill. Analogously to the movement of groundwater, soils and rocks have different permeabilities to different gases. Thus, you might expect gas distribution to result from migrations through permeable materials, moving farther or more quickly through less permeable materials. Gases can become dispersed by passive diffusion processes from areas of higher concentrations to areas of lower concentrations. However, there are often measurable pressure gradients in landfills, and gases will flow from higher to lower pressures.

Once again, we bring up the complexity inherent in the temporal and spatial variation that are likely to occur in complex systems like landfills. Spatially, materials that decompose into gases may be spread unevenly throughout a landfill. The processes of decomposition and other pathways of gas formation do not always occur at the same rate in the same place. There are daily and seasonal shifts in temperature and pressure changes as well as changes in chemical and biological factors shaping the formation of gases. As you read the **Background Information**, think about the kinds of research questions you can develop and how these can be answered by phasing the research activities. We will return to research questions and phases of research later.

Background Information

Landfill gases such as methane, carbon dioxide, and hydrogen sulfide have historically been a problem at this landfill. Concentrations of these gases are often high enough to warrant extravagant safety measures during any sampling work around the monitoring wells. In this research project you will look at which gases are produced by the decomposing waste in the landfill and how those gases migrate from the waste to the surrounding material. While we are concerned with flow of gas rather than of groundwater, some of the factors that are used in discussions of groundwater are applicable to gases as well.

Gas migrates from areas of high pressure to areas of low pressure. An example of this would be when a bicycle tire is punctured: air moves from the tire to the atmosphere, because the pressure inside is greater than the pressure outside. Gases can also move from areas of high concentration to areas of low concentration without any pressure difference through a process called **diffusion**. Any movement of gas through porous materials is controlled by the permeability of the material. Thus, the bicycle tire is made up of rubber which has a very low permeability. Rocks can have a range of gas permeabilities depending on how many connected voids or fractures they contain. Gas moves through rock freely only in places where there is no groundwater. We call this volume of rock above the groundwater surface the **unsaturated zone**.

Review the geology of the area by looking at **Figure 3.9** and studying **Research Note 3.7**.

Research Note 3.7: Review of geological units relevant to Research Problem 4.

Note that there are three different geologic units which occur in the landfill vicinity. The two units that make up the bedrock are both volcanic rocks. The **basalt** is primarily a lava flow, with many cooling fractures along which groundwater or gas can move. Overall, this rock has a moderate permeability to gases. The **silicic debris flow deposits** are made of cemented glassy volcanic particles, mostly coarse-sand to silt size. This rock is strongly cracked and fractured allowing moderate permeabilities in this rock.

A **fault zone** separates the two bedrock units. This feature was formed sometime after the two bedrock units were erupted resulting in the juxtaposition of the two units which you can see in cross section A-A' in **Figure 3.4**. The fault zone is composed of cracked, broken, and jumbled rock fragments, and is thought to have extremely high permeability.

The third geologic unit is the **alluvium**, which by itself forms the shallowest aquifer (see the cross-sections on **Figure 3.1**). This alluvium is a layered sequence of sands, clays, and richly organic layers called "peat." The alluvium has a much higher permeability than the bedrock.

Another unit shown on the map is the landfill. Notice that there are two "types" of waste in this landfill. The waste farther up the valley (north end) is older waste that was primarily dumped from 1960-1985. The landfill surface was sealed off, and the landfill was closed in 1985. The landfill re-opened in 1990, and the waste farther down the valley (south end) is younger waste, deposited from 1990- present. Since the landfill re-opened, its operators have been in compliance with EPA laws that specify cover requirements by using a very impermeable clay-rich soil to seal the landfill surface after major cells have been filled and completed.

In 1989, KRAKO Inc. (a general contracting company) needed some fill material for a nearby housing development. KRAKO began removing portions of the waste from a small area on the northwest corner of the landfill. Using heavy equipment, they moved the waste to the side and then removed the soil developed on the bedrock under the waste. Then they piled the waste back into its original position and covered it. Apparently, the gas levels during this removal were so high that the contractor was unable to continue the work for very long periods of time.

Your job in this laboratory will be to figure out how landfill gas moves from the landfill into the surrounding rock. You'll need to test two hypotheses. We have listed these in the next section, **Research Questions**. Note the kind of language used to state the hypotheses after the research question is posed. If you are not comfortable with the language used to state a hypothesis, review the differences between descriptive and experimental research by reading the relevant sections of the electronic book entitled: **Research Design**.

You should also be familiar with the geologic and groundwater concepts developed in Research **Problems 1-3**, as well as having some knowledge of how bacteria produce gases in decomposing waste. It will also be useful to be familiar with the rudimentary physical properties of gases. Finally, this research problem will require you to use spreadsheet/graphing programs, so that you can enter the data you collect and produce technical figures that show the results of your research.

Research Questions

Gas could move through the landfill through pathways of permeable rock. We need to think about what pathways exist in the landfill, and then design studies that will allow us to determine which pathways are the most likely to allow gas migration. One possibility is the fault zone between the two bedrock wells is acting as a pressurized conduit for gas escaping from the landfill. The gas then escapes from the fault zone into the surrounding bedrock. However, a second possibility is that the KRAKO construction project removed relatively impermeable soil and allowed waste to come into contact with the freshly scraped fractured bedrock. The fractured bedrock could now be acting as a pathway for landfill gas migration. A simplified analysis of the possibilities is provided in **Research Note 3.8**.

> **Research Note 3.8: How many pathways could there be?**
>
> Let's think about the possibilities. Gas might be escaping from the fault zone into the surrounding bedrock and also moving through freshly scraped fractured bedrock which now acts as a pathway for landfill gas. On the other hand, gas might not be moving through either pathway. What are the other possibilities? Well, gas might be moving from the fault zone, but not through the freshly scraped fractured bedrock. Or, the gas could be moving through the freshly scraped fractured bedrock, but not from the fault zone.

Part of the creativity of doing science is to think of all the possible outcomes and their likelihood of occurring. If you have taken a statistics course, you probably learned how to calculate the number of distinct combinations from a certain number of options. In any case, let's get back to this research problem, and try to outline some meaningful hypotheses we can test. We suggest the following approach. You might be interested in **Research Note 3.9** for an overview.

Suppose you use the hypotheses we proposed in **Research Note 3.9**. What reasonable research designs can be used to test these hypotheses? We could measure gas concentrations at each of the available wells on the site. Then, by studying the relative concentrations, we could outline the likely pathways of gas migration. The data supporting the likely pathways of gas migration can be used to evaluate the hypotheses we have proposed.

Your instructor may offer alternatives to the hypotheses we proposed. Again, we emphasize there can be a lot of creativity in scientific research. Our approach is one way, but there are certainly other approaches that can be used. We simply want you to think through the possible options, use your creativity to outline testable hypotheses, and then design research to evaluate these hypotheses. Push yourself to think through the logic of developing hypotheses. If your instructors pose different hypotheses, contrast their approach to ours. By studying the diversity of approaches, you will expand your own creativity by seeing how and why different perspectives can lead to different hypothesis structures and their associated research designs.

> **Research Note 3.9: Possible hypothesis we can test.**
>
> - Develop a hypothesis for each of the two pathways we have outlined above.
> - Write the null hypothesis and alternative hypothesis for each.
> - Design a research project that would allow you to test the hypotheses and reach a conclusion about the most likely patterns of gas migration.
>
> So, the null hypothesis related to the fault pathways could be stated as follows.
>
> > H0: There is no movement of gases from the fault zone into the surrounding bedrock.
>
> An alternative hypothesis could be stated in this way.
>
> > H1: The fault zone between the two bedrock units is acting as a pressurized conduit for gas escaping from the landfill.
>
> For the null hypothesis related to the freshly scraped fractured bedrock acting as a pathway for landfill gas, we could state the following.
>
> > H0: Freshly scraped fractured bedrock does not act as a pathway for landfill gas.
>
> The alternative hypothesis could then be stated as follows.
>
> > H1: Removal of relatively impermeable soil allowed waste to come into contact with the freshly scraped fractured bedrock which now acts as a pathway for landfill gas
>
> You notice we have stated two separate hypotheses. We did not use the fault zone pathway as one hypothesis and the scraped fractured bedrock as its alternative (or the other way around). Spend a little time and critique our approach. Why did we choose to develop two separate hypotheses, each with an alternative? And a very important point for you to address: Are the proposed hypotheses actually testable by a research design that we can develop and execute?

In the next section, we provide phases of research that can be used to test the hypotheses we proposed. Before plunging into the research, study these phases, and determine how each contributes to the collection of data that can be used to test the proposed hypotheses. This is a crucial step, since you as the scientist have to be sure that the questions you want to answer can, in fact, be answered by the data collected through the research process.

Phases Of Research

Phase 1 — Collect gas and groundwater data.

Task 1:
Collect gas and groundwater data using the gas meter and the sonic water level probe at each of the groundwater wells and at the two leachate wells. At each of the leachate wells, you will also see a pressure reading. Record the data on **Table 3.5**, and complete the calculations for groundwater elevation. Notice that the wells in table one are grouped by aquifer.

Task 2:
You'll need a way to evaluate whether the gas is distributed along the fault zone or from the damaged area on the north west corner. One way to do this is to see if there is any discernible relationship between distance from the potential gas source and gas concentration. To collect the distance data you will need a ruler. On the geologic map measure and record the distance from the fault zone to each of the wells in the bedrock aquifers. Your ruler should be perpendicular to the fault zone when you make this measurement. Place a large dot on the northwestern-most corner of the landfill (where the construction work occurred in 1989), then measure and record the distance from the dot to each of the bedrock monitoring wells.

Phase 2 — Build graphic tools for studying relationships between gas concentrations and distance from suspected sources.

Task 1:
To evaluate the two hypotheses, you will need to study the relationship between the gas concentration and the distance from each suspected source. This can be done graphically. Open your favorite spreadsheet-graphing program (EXCEL, Grapher, Quattro Pro,) and enter your gas and concentration data. Use the following headings in your spreadsheet:

Well	Distance to Fault (ft)	Distance to NW Corner (ft)	O_2 (%)	EXP (%)	CO_2 (%)	H_2S (ppm)

Task 2:
You should produce three graphs. For each of the bedrock units plot the wells in that unit versus the distance from the fault. Also, produce a graph of all the bedrock wells versus the distance from the northwest corner. Make sure the horizontal and vertical scales of the graphs are the same so you can easily compare them.

Phase 3 — Synthesis of data and evaluation of hypotheses.

Task 1:
You must answer the overarching question, "How is the gas escaping from the landfill?" We have two hypotheses you can test. Rejection of one hypothesis but not the other will allow you to conclude which pathway is the mostly likely for gas migration. Remember that you might not be able to reject either hypothesis. In this case, neither of the proposed pathways is viable, and you will have to go back to the landfill and rethink the options. Spend some time thinking through the possibilities for gas migration and what kinds of data would support the various pathways.

Task 2:

In **Research Note 3.10**, we provide some guidelines you can use to synthesize your findings into a framework that will allow you to evaluate the hypotheses and make a decision about the most likely pathway of gas migration.

Research Note 3.10: Some guidelines for critical thinking about this research problem.

1. You measured the horizontal distance on the map from the two sources to each well. Strictly speaking, is this the actual distance the gas would need to travel ?

2. Based on your gas measurements, what kinds of bacteria are active in each of the two different parts of the landfill?

3. What about pressure in different parts of the landfill? What does this mean in terms of the way the two parts are sealed from the bedrock and/or the atmosphere?

4. Can you identify which part of the landfill is supplying the landfill gas based on the relative percentages of each gas? This is sometimes referred to as its "signature".

5. How do you explain the concentrations of gases in the alluvium wells? Refer to **Figure 3.10** as you answer this.

In the Know about Where Gases Go

You were the principal scientist responsible for studying gas migration in the landfill. As part of this effort, you are called upon to present your findings to the rest of the scientific team. You are also asked to present your findings to the corporation that operated the landfill. They are considering litigation against KRAKO Inc. if data support the conclusion that the construction opened a pathway for gas migration and, therefore, disrupted the integrity of the landfill. Answer each of the following questions, first for the scientific team and then for non-technical staff from the corporation operating the landfill.

1. What gases are being generated in this landfill, and which gases are dangerous?

2. How and why could gases move through the landfill?

3. Even if the gases that are migrating are not dangerous, do pathways for gas migration create potential problems for the landfill?

4. What are the probabilities that your conclusions are not correct?

5. Why will your research design, data, and conclusions stand the scrutiny of a litigation in which your work is expert testimony?

Table 3.5: A useful data table for Research Problem 4.

Well:	SUE (ft)	DTW (ft)	GWE (ft)	Gas Pressure (psi)	Distance to Fault (ft)	Distance to NW Corner (ft)	O_2 (%)	EXP (%)	CO_2 (mg/L)	H2S (ppm)
MW-1	5418.23									
MW-2	5407.22									
MW-3	5417.89									
MW-5	5387.56									
MW-9	5373.65									
MW-4	5386.29									
MW-6	5350.67									
MW-7	5368.25									
MW-8	5348.29									
MW-10	5348.03									
MW-11s	5347.22									
MW-12s	5371.25									
MW-13s	5350.65									
MW-14s	5368.01									
MW-15s	5351.72									
LMW-1	5393.01									
LMW-2	5383.83									

SUE = Stick-up elevation, the elevation of the top of the well
DTW = The depth to water measured by the meter from the top of the well
GWE = Groundwater elevation = (SUE - DTW)

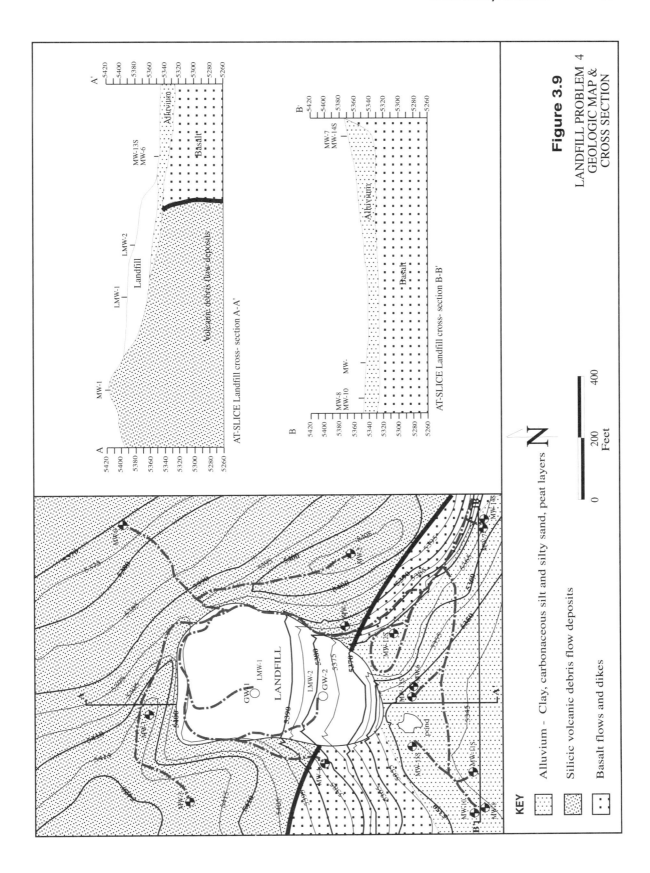

Figure 3.9

LANDFILL PROBLEM 4
GEOLOGIC MAP &
CROSS SECTION

Chapter 3: Land Full of Landfills

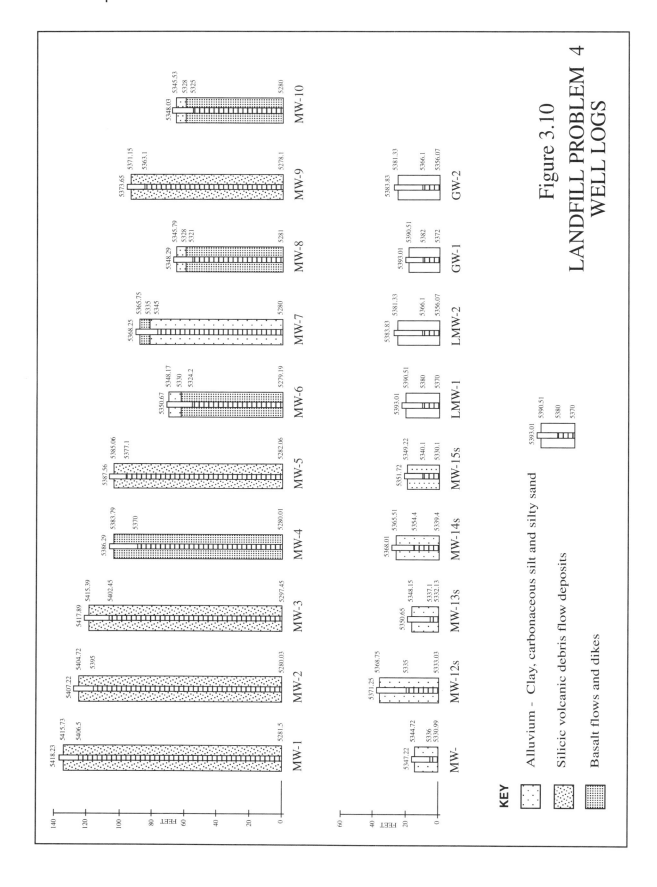

Figure 3.10
LANDFILL PROBLEM 4
WELL LOGS

SO, WHAT'S THE POINT?

This last part of the landfill research focuses on you. We want to get up front and personal by challenging you to extend your knowledge. Here's the challenge.

Take a few days out of your life, and, every time you throw something away, try to figure out where it goes. This may seem simple at first. For example, if you only threw things away at your dorm, apartment, or home, then you could probably quite easily trace when and how the trash was picked up, where it was delivered, and how it was handled at its final resting place. However, you probably travel around on most days, throwing stuff away here and there. Maybe in your city or town all of the solid waste goes to one or a few landfills or incinerators. On the other hand, maybe you contribute each day to waste streams that go to different landfills. The point is to try to figure out where your waste goes.

This process is almost certainly not a simple one. Try to trace the transfer of your waste from garbage can to garbage truck to transfer stations to landfill. Think about what happens to waste thrown away at home, for example, compared to waste thrown away at work or school, or when you are traveling. Identify all of the waste streams to which you contribute. Find out where all of these wastes go, and how well the structural integrity of the landfill is maintained.

Now, calculate the following:
1. Estimate your average daily rate of waste generation in pounds (or kilogram) and in volume.
2. Use the daily rate to estimate monthly and yearly rates of waste generation.
3. Try to estimate amounts of waste your family or friends produce, and compare it to your waste stream.
4. Estimate your city's annual waste production using your average waste as an approximation of per capita waste production.

Take a different approach and figure out:
1. What kinds of waste are you producing?
2. Which of your wastes decompose quickly, and which do not break down easily?
3. Does your lifestyle lead to production of toxic wastes?
4. If you recycle, how has recycling reduced your average waste production?
5. If you don't recycle, what arguments can you use to justify not trying to reuse at least some of the materials you cast off?
6. Compared to other countries, how much solid waste do Americans produce per capita? Why are there differences in amounts of waste produced?

Finally, why bother asking all these questions and trying to answer them?

Book Nook
384-6352
Cerro Coso College

Agency Billing Invoice

Date: 2-23-00
Name: Pat Strick
Account #: 600
Agency: Voc. Rehab

_____

1. _____
2. _____
3. _____
4. _____
5. _____
6. _____

Customer Signature: _Pat M [signature]_
Store Signature: _[signature]_

Notes: _____

```
        390949 CHARGE A    2745 0003 302
        ACCOUNT NUMBER               600
        978013096262 NEW
        KELLY/EXCURSIONS E   MDS 1   38.45
                    SUBTOTAL         38.45
        7.25% SALES TAX               2.79
                    TOTAL            41.24

                    CHARGE           41.24

                              2/23/00  6:29 PM
```

Yucca Mountain – To Store or Not to Store Nuclear Wastes?

WHO CARES WHERE YOU PUT NUCLEAR WASTES?

Are you listening to music while reading this chapter? Or maybe using a light? Electricity is such an integral part of our daily activities that we often take its convenience for granted in the United States. Use of electricity pervades nearly every aspect of our lives, and yet it doesn't seem to cost us all that much. In fact, it takes a fair amount of effort to escape from electricity unless you are used to backpacking into isolated areas. How is the electricity you use generated?

No doubt, you have read or heard about production of electricity from fossil fuels, hydroelectric generators, wind power, solar energy, and nuclear power. You might have even had a chance to seriously discuss the environmental impacts of each type of production. Interestingly, the way power grids are constructed and electrical energy is bought and sold, it might not be easy for you to trace where and how the electricity you use is actually generated. Although there has been a great deal of controversy about the use of nuclear power plants for generating electricity, many Americans receive at least part of their electrical power from nuclear reactors.

You can find out for yourself that most of the high-level nuclear waste in the United States comes from nuclear power plants. Are you surprised that over 4 million pounds of high-level nuclear waste is produced by the nuclear energy industry each year? Currently, many nuclear power plants store the waste materials on site. Is there a nuclear power plant near you, and does it store its own high-level nuclear waste?

The vast majority of Americans use electricity in many activities of daily living. As citizens in a technologically-oriented democracy, we have a responsibility to make informed choices about energy production and usage in the United States. As part of this responsibility, we are now being faced with making decisions about what to do with the waste products of energy production. In the case of nuclear power plants, we have to deal with storage of spent nuclear fuels that are still highly radioactive.

How do you as a citizen answer these questions?

- How is electricity generated by nuclear power plants, and what are the wastes that are created in the process?
- How dangerous are these nuclear wastes? Will the dangers decrease through time? For example, do radioactive materials ever become less radioactive? How can any dangers to humans and the environment be minimized?
- You discover there is a nuclear power plant within 200 miles of where you live. How do you feel about storage of spent nuclear fuels at the power plant?
- Even if we stopped using nuclear power for electricity generation (which is unlikely), we still have to deal with existing nuclear wastes. Are there safe places in the United that can be used to store these wastes?
- Should we choose one place or several locations as repositories for nuclear waste?
- Suppose the government decides to move nuclear wastes from power plants to locations they feel are safe. You look at the United States' road and railway systems and realize that wastes will be traveling through major population centers, including communities in which you have family and friends. How do you feel about transportation of these wastes through or near the communities in which you and your extended family members live?
- You are an advocate for sound environmental management and minimization of electrical power generation that creates dangerous by-products. If alternative methods of clean power generation are available but will be expensive to develop and implement widely,

how will you help teach the American population that their costs for electrical energy are going to increase?
- You think about major catastrophes regularly reported in the news and begin to wonder if there are any places on the face of the earth that are safe storage facilities for dangerous materials like nuclear wastes. After all, what places on the face of the earth are totally free of tectonic activity (possible earthquakes, volcanoes, major earth movements), major storms, erosion processes, floods, changes in groundwater levels, and accidents from human error (like poor construction) or deliberate human action (like terrorism)?
- After all is said and done, what level of risk will you accept? For example, would you accept the risk of an accident with nuclear wastes that is the same as your risk of you being involved in a car accident?

WHAT DO YOU SEE IN THE REAL WORLD?

One solution to the accumulation of nuclear wastes from electricity generation is to store them in a safe place. What do we mean by a safe place? Well, in these times our definition of a safe place is a location with characteristics that minimize the risks to humans. To minimize risks, a storage location would have to be constructed so that radioactive materials do not leak out, no matter what happens. What could happen? There are certainly natural disasters like earthquakes, volcanoes, floods, tornadoes, and hurricanes. Plus, there are natural processes that are not usually considered disasters like changes in precipitation and groundwater level, wind and water erosion, physical and chemical breakdown of materials from sunlight and weathering, and patterns of global climate change. Don't forget human problems. Human error can lead to accidents in construction of storage containers and facilities, during transportation of nuclear materials to storage sites, and in the maintenance of sites. Unfortunately, there are some risks from terrorists. Perhaps the greatest danger of all is basic ignorance about how much we need to know in order to make sound decisions about the management of complex environmental and human systems.

We have to deal with the accumulation of high-level nuclear waste materials, and even if we stopped using nuclear power plants tomorrow, we would still have the problem of the wastes already produced. One plan now under consideration is to store this waste, along with high-level nuclear waste from several nuclear power plants from Pacific nations, in a proposed nuclear waste repository at Yucca Mountain. This site is located in Nevada, northwest of Las Vegas. Nuclear waste would be transported to Yucca Mountain from storage locations at power plants or other repositories. As you might guess, the use of highway and railroad transportation would result in nuclear waste passing through or very close to a large number of American cities and towns on its way to Yucca Mountain.

The Yucca Mountain site was originally proposed because people thought it met a number of safety criteria. If you think about safety, there are two broad areas of concern. One is the stability and maintenance of the site itself. The second concern is how to get nuclear wastes to the site safely.

In regards to stability and maintenance of the site, the nature of nuclear wastes will require that any site be designed to hold nuclear waste for time periods of tens of thousands of years. That's a long time. Such a time scale is pretty much outside our imagination, since no one knows of any buildings that have maintained their structural integrity for tens of thousands of years. When we talk about tens of thousands of years, we are in the realm of the "geologic time scale," Yucca Mountain was proposed as a "geologic repository," which means that the geologic features were expected to be sufficiently stable so that nuclear wastes could be contained safely for a very, very long period of time.

As for safety during transportation, there are a number of problems. How do we store the materials during transportation? How will the materials be moved across the country from power plants to the Yucca Mountain site? How many shipments will be needed? Given the method of transportation, the type of packaging, and the number of shipments, how do we calculate the probability of an accident, and, of these accidents, how many are likely to pose risks to the environment and humans? Of course, these questions are only the beginning. As soon as you try to answer these questions, new questions will be generated.

For example, the Department of Energy has estimated a repository would receive about 6,200 truck shipments and 9,400 rail cask shipments of spent nuclear power plant fuel and other radioactive wastes. Because of the locations of the power plants and existing storage facilities, shipments to Yucca Mountain would pass through a number of cities. The exact number and location of cities will depend on the particular choices of routes and modes of transportation (State of Nevada Nuclear Waste Project Office, May 17, 1998). Choices of routes and modes of transportation will not be made only on a scientific basis. You can imagine that many communities will not want the waste to pass through or near them. Talk about politics! We will be entering a period of intense discussion about the choice of routes, and the outcomes will be the result of complex social, political, historical, economic, and scientific arguments.

So, the main point is that you are one of the scientists who have to make some decisions about Yucca Mountain and its use as a repository for nuclear wastes. Your research has to be as rigorous as possible. People's lives will depend on your conclusions. You have to take a long-term view. Will your conclusions stand over a period of tens of thousands of years? Do not forget our discussions of temporal and spatial heterogeneity when we worked on the landfill research. You might be able to control the spatial variability by choosing a limited site like Yucca Mountain and using a combination of engineering and geology to create what you believe is a stable site. However, think of the variability of the real world over time, especially over tens of thousands of years. Your problem is simply to anticipate every variable that might influence the site, to set priorities for the importance of these variables, and to model their possible impacts on the site. Good luck!

DEALING WITH COMPLEXITY

You have probably gotten the hint by now. Your research will focus on familiarizing yourself with the proposed high-level nuclear waste repository at Yucca Mountain. This site is located on the Nevada Test Site and Nellis Air Force Bombing Range northwest of Las Vegas, Nevada. We suggest you spend time in the Virtual Office *Bookshelf*, and read the books related to research on Yucca Mountain. These will give you some background that will help you understand the proposed repository, why many people think it is a good and necessary choice, what materials are expected to be stored there, and what potential effects it might have on the area.

You must assume a variety of roles, from government regulator to research scientist. In these roles, your studies will be designed to gather the information needed to make a preliminary determination about whether Yucca Mountain is a suitable site for a high-level nuclear waste repository. You will have four research opportunities.

- **Nuclear Waste Repository Research Problem 1:** Yucca Mountain – A Reasonable Site? 89
- **Nuclear Waste Repository Research Problem 2:** Tectonic Activity and Repository Site Stability . .101
- **Nuclear Waste Repository Research Problem 3:** Transportation of Nuclear Wastes 115
- **Nuclear Waste Repository Research Problem 4:** Risk of Volcanic Activity 125

Nuclear Waste Repository Research Problem 1: Yucca Mountain – A Reasonable Site?

Introduction and Advice

Before you start the research, examine **Study Notes 4.1** and **4.2** for some guidance about what you need to know before you start and what you should know when you are done. In **Study Note 4.3**, we provide a table you can use to keep track of assignments and their due dates.

> **Study Note 4.1: As you begin the research, you will need to have some prior knowledge in the following areas:**
>
> 1. A basic understanding of radioactivity.
> 2. Introductory knowledge about the nuclear waste disposal problems of the United States.
> 3. Some knowledge about the complex political and sociological issues associated with siting a high-level nuclear waste repository.

> **Study Note 4.2: What you should be able to do after completing the research.**
>
> 1. Investigate geologic formations that will determine whether the proposed repository site is adequate.
> 2. Understand the groundwater systems in the repository area and their likely impact on the site.
> 3. Study seismic activity in the Yucca Mountain area, and determine if there are active faults near the repository site.
> 4. Search for evidence of volcanism in the repository region.
> 5. Identify risks to the human populations near the repository site.
> 6. Predict risks to and from the site over a long period of time.

Study Note 4.3: Use the table provided below to list the research assignments given by your instructor. Make notes on this table about the types of work you will be expected to submit for a grade.

Research Assignments	Specific Instructions for Completing Assignment	Completion Date
Phase 1		
Phase 2		
Phase 3		
Phase 4		
Phase 5		
Phase 6		
Other Assignments		

What's the Problem?

Here's the problem. We have quite a load of nuclear waste materials generated by nuclear power plants. There are also some nuclear wastes that resulted from production of nuclear materials. These wastes are radioactive, and they pose danger to the human and earth ecosystems. We need to put them in a safe place.

Our first problem leads us to a second problem, which is setting the criteria for a location in order to declare if it is *safe* for storage of nuclear waste. We would like these criteria to be so stringent we can say almost with certainty that *no matter what happens* nuclear wastes will not get out of the repository and damage the environment. Each proposed site we study will have to be examined in the context of these safety criteria. Those that pass can be considered as potential repositories for nuclear wastes.

Fine, but how do we start setting safety criteria? Well, first we make a complete list of all the disturbances and human errors that might cause stored nuclear wastes to get into the environment. Then we search for or create that combination of an engineering and geological site that will allow us to say with near certainty, "No nuclear waste will escape into the environment from this site."

Our *certainty* in this matter must be an absolute confidence in our conclusion that the safety of the storage site can pass a severe test of time. We are not talking about just a few decades, or even a few centuries. Basically, the radioactive wastes from nuclear power plants decay from dangerous materials to not so dangerous materials over very, very long periods of time. So, when we finally stop putting wastes in the repository and close the site, we will still have to wait tens of thousands of years for the dangerous levels of radioactive materials to diminish to not so dangerous levels of radioactivity. During this time, the site must be maintained and free of natural and human-induced disturbances that could release radioactive materials into the environment.

Don't be overwhelmed. Someone has to study this problem, and that person should be someone you trust. Who better than you to be part of the scientific team who will study potential repositories for their suitability as long-term storage sites of nuclear wastes? You know from the landfill research that we recommend organizing your research into phases. One strategy for trying to study the suitability of any repository site is to conduct research in six phases. Please note as you read the phases we list below that they are not independent of each other. That is, the data you gather in one phase may be important to data collection, analysis, and interpretation in another phase. Also, work with your classmates to see if you can see a different way to structure the research into different phases of data gathering and analysis.

1. Study the geological formations in the repository area, and find out how they formed and how stable they are.
2. Examine the groundwater in the repository area. Will changes in groundwater levels cause problems for the repository?
3. Review the data on seismic activity, and determine if there any active faults that cut through the proposed repository site.
4. As you study the proposed site, search for evidence of volcanism.
5. Study the human demography of the proposed site and the region in which the site is located. Can you identify unacceptable risks to these people?
6. Use the data from the five research efforts listed above to predict the risks to and from the site over a long period of time.

92   Chapter 4: Yucca Mountain

Read the **Background Information**, and then head for Yucca Mountain. There's work to be done, and, unfortunately, you don't have tens of thousands of years to complete the research.

Background Information

Now, to get to some specifics. Your research will focus on a proposed high-level nuclear waste repository at Yucca Mountain, Nevada. This site is located on the Nevada Test Site and Nellis Air Force Bombing Range northwest of Las Vegas, Nevada. In this research simulation, you will spend much of your time familiarizing yourself with the proposed site. To start, you will need to read the books in the *Bookshelf* that are related to Yucca Mountain. These will give you some background on the proposed repository, why it is deemed necessary by many people, what materials are expected to be stored there, and what potential effects it might have on the area.

You play the role of a government regulator in this first Yucca Mountain research activity. Your responsibility will be to gather together the information needed to make a preliminary determination of whether Yucca Mountain is a suitable site for a high-level nuclear waste repository. This information comes from scientific research, engineering designs, and political considerations. Your job is to integrate this information and develop a complete sense of what parts of the research results favor the approval of the site for a repository, and what parts might require further research.

First, explore the Yucca Mountain simulation. You will find a **Question and Answer room**, a **Theater room**, a **Map room**, and a **Transportation room**. The mode of presentating information in each of these rooms is distinct. Go to each room, and see the types of information available. *The majority of this assignment will ask you to gather information by visiting these various rooms.* If you are familiar with them in advance, it will allow you to get the necessary information more quickly, and, since you may need to visit two or more rooms to answer a particular question, you will know when you have gathered all the available information to answer the question.

The proposed nuclear waste repository at Yucca Mountain will be designed to hold nuclear waste for time periods of tens of thousands of years. This time scale is far beyond that of usual human activities and the structures they build. For example, what country has a continuous history longer than 5,000 years? Tens of thousands of years is in the realm of the "geologic time scale." Because of this, the repository must be more than simply a carefully designed building, since no one knows of any buildings that have maintained their structural integrity for tens of thousands of years. Instead, the plan is to use geologic rocks and structures to hold the nuclear waste for such great lengths of time. In order for this plan to work, the geology must be proven suitable to allow the waste to remain undisturbed. A tremendous amount of study has now geologically characterized the Yucca Mountain area better than probably any other area in the United States.

In the **Repository Requirements** book, you will find some of the requirements for a nuclear waste repository. The studies at Yucca Mountain have been designed to determine whether the site meets, or can meet, these requirements. Fundamentally, we must know whether the waste can be isolated from the surface environment for a sufficient length of time. This means that water will

not carry the waste to the surface, volcanoes will not erupt through the site and spew the waste into the atmosphere, and earthquakes will not rupture the site so the waste can be carried to the surface by water.

The questions you will answer in this laboratory are driven by the requirements for a nuclear repository, as required by the Nuclear Regulatory Commission and reviewed in the **Repository Requirements** book.

In regards to your prior knowledge, you must have some basic background in geology to conduct this research. For example, you have to know what rocks are, how they are deposited, and have an introductory knowledge about geological formations. You might also want to review information on tectonic activity, especially earthquakes and volcanoes.

Research Questions

We have said that the general problem is deciding if Yucca Mountain meets the requirements for being used as a high-level nuclear waste repository. The six phases of research provide one framework for collecting and analyzing data so we can decide if the Yucca Mountain site meets repository requirements. Try to identify a specific research question for each phase. Use this question to provide a focus for your efforts.

Look again at the six phases we proposed.

1. Study the geological formations in the repository area, and find out how they formed and how stable they are.
2. Examine the groundwater systems distributed in the repository area. Will these systems cause problems for the repository?
3. Review the data on seismic activity and determine if there are any active faults that cut through the proposed repository site.
4. As you study the proposed site area, search for evidence of volcanism.
5. Study the human demography of the proposed site and the region in which the site is located. Can you identify unacceptable risks to these people?
6. Use the data from the five research efforts listed above to predict the risks to and from the site over a long period.

For **Phase 1**, we might pose the following question:

What are the geological formations in the Yucca Mountain area? How did they form?

In **Phase 2**, how about this research question:

Are there any active faults that cut through the proposed repository site?

Get the idea? Look at each phase, and set up a question or set of questions which will become a guiding framework for your research. Take a look at the phases of research listed below. You will see we have provided some guiding questions you can use to direct your review of the existing data on Yucca Mountain. Read through the phases and their guiding questions, then use the simulation and the *Bookshelf* to review data from earlier studies.

Phases of Research

Phase 1 — Studying the geological formations in the repository area.

In this first phase, we are interested in the types of rock formations. We also want to know how they formed, and how stable they are. (You will find that many of the answers to these research questions will be found by combining the data from the simulation —especially the maps in the **Map room**— with the reference materials on the office bookshelf.)

Task 1:
Examine the rock types in the Yucca Mountain area. What types do you find, and how did they form?

Task 2:
Now, describe the hydrologic properties (the properties that control water flow and permeablility) of the rocks at Yucca Mountain.

Task 3:
Decide which rock units might make good 'roofs' for a repository, and prevent water from gaining access to the waste. What properties are you looking for to answer this?

Task 4:
Which rock units might serve to slow the spread of nuclear waste if it did become dissolved in water?

Phase 2 — Examine the groundwater systems distributed in the repository area.

We can now turn to studies of the groundwater. We need to know if groundwater could cause problems for the repository.

Task 1:
Try to figure out where the groundwater goes after it leaves the Yucca Mountain area. Where does it come to the surface, and how does it get there?

Task 2:
Assess whether there is a problem if contaminated groundwater comes to the surface. How might nuclear waste spread from the groundwater discharge site?

Task 3:
How long would groundwater take to get to the discharge site? How much lower would the radioactivity be when it gets there?

Phase 3 — Review the data on seismic activity of the area.

In this phase, we will study the Yucca Mountain area, and examine the possibilities for active faults that might cut through the proposed repository site.

Task 1:
Determine if there are any active faults that cut through the proposed repository site.

_____
_____
_____
_____

Task 2:
Decide if there are any active faults near the site. This requires that you decide how far away is still 'near'.

_____
_____
_____
_____

Task 3:
Based on your knowledge of the size and proximity of earthquakes likely in the area (the bookshelf mighty come in handy), how bad would you say the earthquake hazards are at the site? In order to answer this, think about which areas tend to shake most in an earthquake. Where would you expect more shaking: in the repository, which is carved in bedrock, or at a processing center on gravels at the entrance?

_____
_____
_____
_____

Phase 4 — Is there any evidence of volcanism in the Yucca Mountain area?

Another potential hazard at Yucca Mountain is active volcanism. The area is in a field of diffusely distributed cinder cones, and an eruption through the site could potentially spew nuclear waste into the atmosphere. If the vents have an essentially random distribution, the risk of an eruption at the Yucca Mountain site is the same as anywhere else in the area. If there is some sort of control on their distribution, volcanic risks will vary from place to place. In some cinder cone fields, vents are found along lineaments that correspond to orientations of faults in the area. In such cases, the greater hazard is in places that lie along that lineation.

Task 1:
Look at the distribution of the cinder cones in Crater Flat. Do you see any possible lineaments? If so, what is their orientation? Draw the possible lineaments on the attached map (**Figure 4.1**). Does Yucca Mountain lie along one of these lineaments?

_____
_____
_____
_____
_____
_____

Phase 5 — Study the human demography of the proposed site and the region in which the site is located.

We will need to ascertain if there are any unacceptable risks to people living near the Yucca Mountain site. Part of this effort involves identifying potential risks. However, we also have to assign a likelihood that each risk will occur. Review the book: Yucca Mountain Story.

Task 1:
Determine the current population around Yucca Mountain. Approximately how many people live within a 100 mile radius?

_____
_____
_____
_____

Task 2:
Estimate the predictions for future population growth in the Yucca Mountain area. You will need to define your area boundaries to answer this question.

_____
_____
_____
_____

Task 3:
Describe how the Yucca Mountain area has been used by people in the past.

_____
_____
_____
_____
_____

Chapter 4: Yucca Mountain

Phase 6 — Use the data from the five research efforts listed above to predict the risks to and from the site over a long period of time.

All of the above questions ask you to use the present conditions at Yucca Mountain to predict the risk in the future. For earthquakes and volcanoes, there is some use of past evidence (how big earthquakes have been and where volcanoes have erupted). For groundwater, the assumption is that present climatic conditions will continue into the future. This assumption, however, must be investigated. We live in a time period of relative climate stability, a period that has extended since the end of the last glacial period about 10,000 years ago. Prior to 10,000 years ago, the climate was considerably colder and wetter in many parts of the globe. If such a global climate change occurred in the future, could it affect the climate in the Yucca Mountain area, and perhaps the groundwater flow and water table height?

One way that scientists have approached this question is by doing the obvious: looking at what happened last time there was glacial period. They then predict what would happen in the future based on the past. First, they must determine what the climate was like in the Yucca Mountain area during the last Ice Age. Did it follow the overall worldwide colder and wetter trend, or was it drier and/or warmer, as occurred at some sites?

Task 1:
Look at the information on past climate in the Yucca Mountain area (see the book: Yucca Mountain Climate). What was the climate at Yucca Mountain 18,000 years ago, at the height of the last glacial advance?

Task 2:
Now, what are some of the ideas proposed about the effects of a climate change on the rate of groundwater flow and the height of the water table. Would the repository be flooded?

Task 3:
Go back to the book: **Repository Requirements**. Based on what you have learned, does the site meet the requirements for a high-level nuclear waste repository? What unsolved problems still remain?

_____
_____
_____
_____
_____
_____
_____
_____
_____
_____
_____
_____
_____

In the Know about Yucca Mountain Site Feasibility

You were responsible for studying some of the repository requirements for storing nuclear wastes. Your research focused on Yucca Mountain. As part of the scientific team, you have been called on to present your findings at a professional scientific meeting and, also, to a press conference attended by reporters from newspapers, radio, and television. Answer each of the following questions, first for the attendees at the professional meeting and then for the news media.

1. What are the nuclear waste disposal problems in the United States?

2. Why was Yucca Mountain considered as a possible site for a nuclear waste repository?

3. What are high-level nuclear wastes and why are they dangerous?

4. Is it safe to move waste to Yucca Mountain ?

5. Is Yucca Mountain a safe repository for the United States' nuclear waste ?

6. What effect will closing Yucca Mountain for more than 10,000 years have on people living near the site as well as people in region?

7. Will the groundwater be affected by the storage of nuclear waste at Yucca Mountain?

8. Are the risks associated with Yucca Mountain acceptable ?

9. What are the scientific, political, and sociological issues associated with siting a high-level nuclear waste repository?

100 Chapter 4: Yucca Mountain

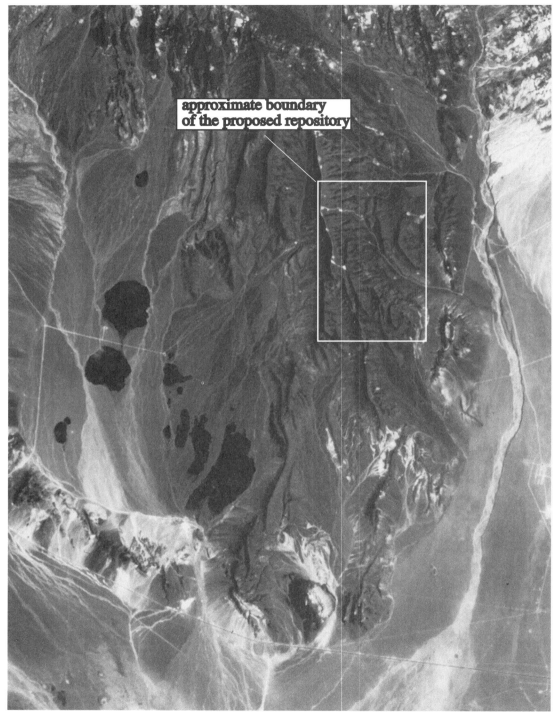

photo courtesy of the Yucca Mountain Project

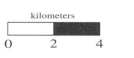

YUCCA MOUNTAIN PROBLEM 1
LINEATION LOCATION MAP

Figure 4.1

Nuclear Waste Repository Research Problem 2: Tectonic Activity and Repository Site Stability

Introduction and Advice

Before you start the research, examine **Study Notes 4.4** and **4.5** for some guidance about what you need to know before you start and what you should know when you are done. In **Study Note 4.6**, we provide a table you can use to keep track of assignments and their due dates.

> **Study Note 4.4: As you begin the research, you will need to have some prior knowledge in the following areas:**
>
> 1. An understanding of the requirements for a high-level nuclear waste repository.
> 2. An understanding of the problems and attributes of Yucca Mountain as a high-level nuclear waste repository.
> 3. Knowledge about the political and sociological problems associated with siting a high-level nuclear waste repository.

> **Study Note 4.5: What you should be able to do after completing the research.**
>
> 1. Determine the requirements for a high-level nuclear waste repository.
> 2. Evaluate the potential for problems due to volcanism, earthquakes, and groundwater at Yucca Mountain.
> 3. Evaluate what information is still needed to decide whether the Yucca Mountain site is appropriate.

Study Note 4.6: Use the table provided below to list the research assignments given by your instructor. Make notes on this table about the types of work you will be expected to submit for a grade.

Research Assignments	Specific Instructions for Completing Assignment	Completion Date
Phase 1		
Phase 2		
Phase 3		
Phase 4		
Other Assignments		

What's the Problem?

We were introduced to the Yucca Mountain area in **Research Problem 1**. Let's revisit some important points about the site. First, detailed requirements for a high-level nuclear waste repository are given in the **CFR (Code of Federal Regulations) Books**. Among the requirements is that the repository must be sited and constructed in such a way that it contains the nuclear wastes and protects the environment from dangerous levels of radiation. This protection must last for at least 10,000 years after the last wastes are deposited and the repository is closed. We mentioned earlier, and repeat emphatically here, that tens of thousands of years represents a time longer than any of the known continuous civilizations on earth.

Even if the United States lasts tens of thousands of years, we need to have some idea about what is likely to happen to the site. Forget sociological and political changes for the moment, and just focus on natural process that change the shape of earth. We mentioned tectonic activity, weathering and erosional processes, seasonal changes in precipitation and potential impacts on groundwater, and global changes in climate that might have a variety of impacts on the geomorphology of a region. How do we study the likely events of the future?

In **Research Problem 1**, we began some preliminary work on the geology of the Yucca Mountain area. We also looked at groundwater and examined the area for faults with seismic activity. You recall that another area of interest was the likelihood of a volcanic eruption. It is time to look at some of these potential problems in more detail. For **Research Problem 2**, we are not going to suggest phases of research. Instead, we ask you to examine the **Background Information** presented below, and then use that as a context in which to critique the focus for research we suggest in the Research Questions section.

Background Information

For this research problem, you will examine the important features of the Yucca Mountain area being reviewed by a number of government agencies to determine whether the site is acceptable as a national high-level nuclear waste repository. As you learn the material, you may find there are a number of terms unfamiliar to you. If you start getting confused, go back to the books in the *Bookshelf* dealing with topics in nuclear physics and chemistry, and review the material.

Your workspace for this laboratory is an office with four rooms accessed directly from the office. These rooms are the **Transportation room, Question and Answer room, Theater room**, and **Map room**. Each of these rooms has a different way of presenting information, so that some information, due to its very nature, is best presented in one room and not others. In other cases, similar information is available in several rooms, but you may find that you understand the material better in one presentation style.

Yucca Mountain was originally proposed as a "geologic repository," which means that the geologic features were expected to be sufficient to contain the nuclear waste for the requisite time period. Ideas have changed a bit over time, such that it is now expected to be a "geological and engineered repository." This means that minor problems can be solved through careful engineering, but it must still fundamentally be a geologically suitable site. Therefore, we will start by looking at the geological suitability of the site.

So, what are the basic requirements? Detailed requirements for the high-level nuclear waste repository are given in the **Repository Requirements** and the **CFR** Books. In simplest terms, the repository must protect the environment from dangerous levels of radiation for at least 10,000 years after the repository is closed to any new additions of waste. Some important factors in deciding whether this will occur include the likelihood of a volcanic eruption or earthquake, and whether ground water is likely to inundate the site. In addition, the nuclear waste must be transported from its current location at nuclear power plants around the country to the nuclear waste repository. This requires that the transportation mode be safe and secure.

As always, we want to make a few suggestions about the knowledge you need before entering this research problem. First, you need to know a little bit about the nuclear waste disposal problems facing the US, including the characteristics of high-level nuclear wastes and their decay, as well as the dangers of radioactivity to the environment and human physiological systems. Second, you should know something about the Yucca Mountain Project. If you have completed the first research problem on Yucca Mountain, you will have a basic understanding of the site. If you have not completed this problem, you might want to review some of the electronic books in the *Bookshelf* that are related to Yucca Mountain.

Now, stop for a moment and write down some problem areas that should be studied as we move towards evaluating Yucca Mountain as a high-level nuclear waste repository. For each problem area, try to jot down a few focused research questions you would like answered. Use the space provided below.

Problems that Need to Be Studied	Research Questions

Research Questions

After reading the **Background Information**, we think the following research questions emerge.

1. Is there an overview or synthesis of available data on the site?
2. Is a volcano likely to erupt through the site and send nuclear waste into the atmosphere or cause the groundwater to rise and inundate the site?
3. Are there any faults that might move within the repository itself? Are there are any faults

nearby capable of producing large earthquakes that might cause repository walls to collapse or create changes in the water table so that ground water inundates the repository?
4. What is the likelihood of problems with groundwater intrusion into the site? For example, what is the groundwater contour surface? Will groundwater flow patterns disqualify the site because wastes could not be stored more than 200 meters below the surface?

How do these compare to your problem areas and their associated questions? You note that we did not list problem areas and then research questions for each area. Part of doing science in collaboration with members of a research team involves recognizing and valuing differences in the way a research problem is approached. In earlier laboratories, we structured the research so you could pretty much follow a problem outline with phases of research and attach a specific research question to each phase. Now, we are asking you to contrast and compare the research questions you have developed with those we have developed. Try taking our questions and fitting them into the problem areas you listed above. Also, look at the wording of our questions, and think about how and why they might be different from the wording of your questions.

In the next section of this chapter, we suggest a sequence for conducting the research for this problem. This sequence was based on our experience with research in such systems. Before starting the simulation, read through our phases and critique them. Think of ways to do the research in a different order and evaluate the pros and cons of different sequences of research activities. We don't claim to know the truth. The sequence we chose is based on our experience. However, you are part of the scientific team. Don't just follow our lead. Think deeply about why we developed this particular sequence and why other sequences might be just as good or better. Remember that the lives of humans and the integrity of the environment may depend on the conclusions we reach in our research.

Phases of Research

Phase 1 — Becoming familiar with Yucca Mountain.

In this first phase, we need to explore the Yucca Mountain site and become familiar with the repository.

Task 1:
Enter the virtual environment of the Yucca Mountain simulation and explore the different rooms. Read the books in the Office related to Yucca Mountain. Think about where you would go to find information on particular subjects. Review the electronic book entitled: **Repository Requirements**, and the legal description of these requirements as excerpted from the Code of Federal Regulations in the **CFR** Books. You are now ready to begin the work for this lab.

Phase 2 — Likelihood of volcanic activity.

The concern here is whether a volcano is likely to erupt through the site and send nuclear waste into the atmosphere, or cause groundwater to rise and inundate the site.

Chapter 4: Yucca Mountain

Task 1:
Review the types of volcanoes at Yucca Mountain. Note that the site was built by one type of volcano called a caldera, and that another type of volcanism, basaltic cinder cone activity, is now present in the area. Look at the geologic map in the **Map room**. Concider the eruptive history of the area and think about the idea of the Zone of Most Recent Volcanism as outlined in the **Volcanic Risks** book. Do you see any possible alignments of volcanic vents? Describe them, and draw them on **Figure 4.2.**

Task 2:
Determine where cinder cones have erupted in the past.

Task 3:
Decide where you think basaltic eruptions are most likely to occur in the next 10,000 years, and explain why.

Task 4:
Think about what else you need to know in order to evaluate whether your decision is correct.

Task 5:
If a cinder cone erupted through the repository, the effects on the nuclear waste could vary. Why might these effects vary? Does this increase or decrease the chance of eruption-induced release of radionuclides into the environment?

Phase 3 — Likelihood of earthquake activity.

Our concern about earthquakes focuses on whether there are any faults that might move within the repository itself, and whether there are any faults nearby capable of producing large earthquakes that might cause: (1) wall collapse within the repository or (2) a rise in the water table so that groundwater inundates the repository.

Look at the faults map of the area in the **Map room**, and note where faults in the Yucca Mountain area have been active in the last 11,000 years. Note that many of the faults have not been active in the past 11,000 years, and many of these faults have likely had little activity in the past 10 million years.

Task 1:
Decide how big the earthquakes have been, and estimate the distance of their faults from the repository.

_____
_____
_____
_____
_____

Task 2:
There is a direct relationship between the amount of displacement along a fault (how far it moves) and the magnitude of the earthquake. You can estimate the maximum earthquake magnitude possible along a fault by estimating the maximum rupture dimensions of the fault and, using empirical correlations, calculating the magnitudes associated with those dimensions. The empirically derived equation for this relation is Magnitude = 5.08 + (1.16 x log [rupture length]) (Wells and Coppersmith, 1994). Faults typically consist of segments that rupture as one piece during an earthquake. Thus, the rupture length may not be the same as the total length of the fault.

Select three active faults shown on the fault map and measure their length. Mark the faults you are measuring on the attached map, **Figure 4.3**. Assume in this case that the fault lengths you see are one segment. (Is this a good assumption?) Calculate the maximum earthquake magnitude for these faults, and write it on the map next to the fault. Show your calculations below.

_____
_____
_____
_____
_____

Task 3:

Earthquake magnitude is a measure of the amount of energy released by an earthquake. The amount of earth shaking in any given location is related to the earthquake magnitude and proximity, as well as the type of underlying geological material. For example, shaking is much greater in areas on soft sediments, soils, or muds than it is in sites that are on bedrock. In a tunnel through bedrock, shaking from any given earthquake is likely to be at its minimum. In a magnitude 5.4 earthquake in 1992 at Little Skull Mountain about 20 kilometers southeast of the ESF (Exploratory Studies Facility), shaking was insufficient to cause objects to fall off of small ledges in the tunnel.

Given this information, do you expect ground shaking to be a significant problem at Yucca Mountain? Why?

_____
_____
_____

Task 4:

If an earthquake occured on a fault that cuts through the repository, it could do much more damage by destroying waste containers or allowing water to gain access to the site. Another direct relationship exists between earthquake magnitude and maximum displacement (maximum movement during a single earthquake along the fault; Wells and Coppersmith, 1994). We care about maximum displacement, because this would affect how much movement there might be on the Bow Ridge Fault, an active fault that crosses the proposed repository entrance tunnel. This would tell us how much offset there might be in the tunnel. The relationship is Magnitude = 6.69 +(0.74 x log [maximum displacement]).

Calculate the maximum displacement likely along the Bow Ridge Fault that cuts through the north entrance tunnel of the ESF. Show your calculations below, and write your displacement answer next to the Bow Ridge Fault on your map (**Figure 4.3**).

_____
_____
_____

Task 5:

Now, figure out what potential problems a significant displacement would pose for the repository if the earthquake occurred during the period when nuclear waste was being deposited in Yucca Mountain.

_____
_____
_____
_____
_____

Task 6:
Describe potential problems such a displacement would pose for the repository if the earthquake occurred during the period after the repository has been closed and sealed.

Task 7:
Make a list of other data you need in order to decide if the site is safe.

Phase 4 — Likelihood of problems with groundwater.

Groundwater is the most studied part of Yucca Mountain. This is because water would be the most efficient transport mechanism to move the radioactivity into the environment, both below and above ground. As a first step, the site should satisfy the favorable condition that the waste be greater than 300 m below the surface. The site would be disqualified if the waste could not be stored more than 200 m below the surface. The repository should also be in the unsaturated zone, with the water table far below the host rock so that none of the host rock is saturated.

Task 1:
To determine whether the site meets these conditions, you must first figure out where the water table is located. This requires that you draw a potentiometric surface map for the groundwater. In the **Map room**, find the overlay on the Groundwater map that gives the well data. Each of these wells has been tested for the depth to the groundwater. These well sites are located on the map attached to this laboratory (**Figure 4.4**). Plot the depths to groundwater at each well, and then draw potentiometric contour lines around them. Colored pencils will help distinguish between your lines, and make it easier for you (and your instructor) to read and interpret the map. In order to make the water table's contours easily interpretable, use the following contour elevations: 1000 m, 900 m, 800 m, 775 m, 750 m, 740 m, 730 m, and 720 m.

Do you have any concerns about groundwater after building this map?

Task 2:

Now look at the cross-section of the rocks of Yucca Mountain attached to this laboratory (**Figure 4.5**). Plot the elevation of the water table on the cross-section. Where is the water table with respect to the proposed repository location? How far below the location is the water table? Is any part of the host rock (the Topopah Spring Ignimbrite) saturated?

_____
_____
_____
_____
_____
_____

References

Wells, DL, and Coppersmith, KJ, 1994, New empirical relationships among magnitude, rupture length, rupture width, rupture area, and surface displacement, Bulletin of the Seismological Society of America, v. 84, p. 974-1002.

In the Know about Tectonic Activity near Yucca Mountain

Sometimes, we change our ideas about a complex system after we have had a chance to study it more thoroughly. Suppose that you are once again asked to present your findings at a professional scientific meeting and, also, at a press conference attended by representatives from various communications and news media (i.e., newspapers, radio, and television). Take each of the questions you answered in **Research Problem 1** and evaluate the responses you gave to the attendees at the professional meeting and then to the news media. Since you now know more, have any of your responses changed? The questions are listed below for your convenience.

1. What are the nuclear waste disposal problems in the United States?

2. Why was Yucca Mountain considered as a possible site for a nuclear waste repository?

3. What are high-level nuclear wastes, and why are they dangerous?

4. Is it safe to move waste to Yucca Mountain?

5. Is Yucca Mountain a safe repository for the United States' nuclear waste?

6. What effect will closing Yucca Mountain for more than 10,000 years have on people living near the site as well as people in region?

7. Will the groundwater be affected by the storage of nuclear waste at Yucca Mountain?

8. Are the risks associated with Yucca Mountain acceptable?

9. What are the scientific, political, and sociological issues associated with siting a high-level nuclear waste repository?

Virtual Reality Excursions 111

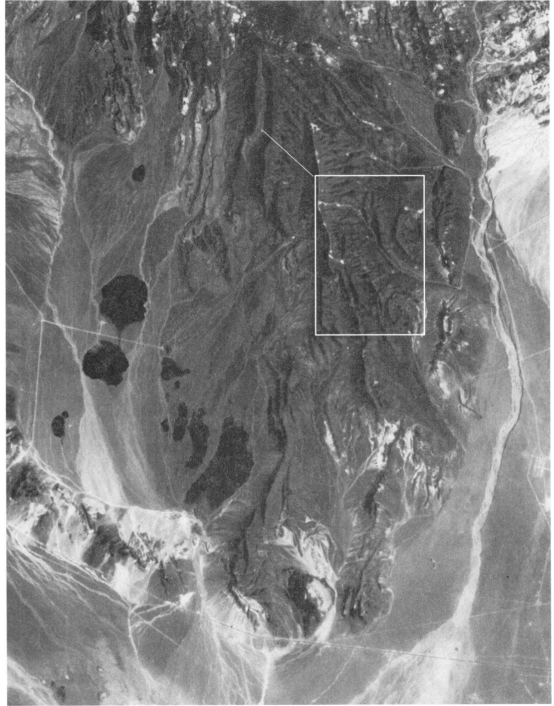

photo courtesy of the Yucca Mountain Project

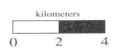

YUCCA MOUNTAIN PROBLEM 2
LINEATION LOCATION MAP

Figure 4.2

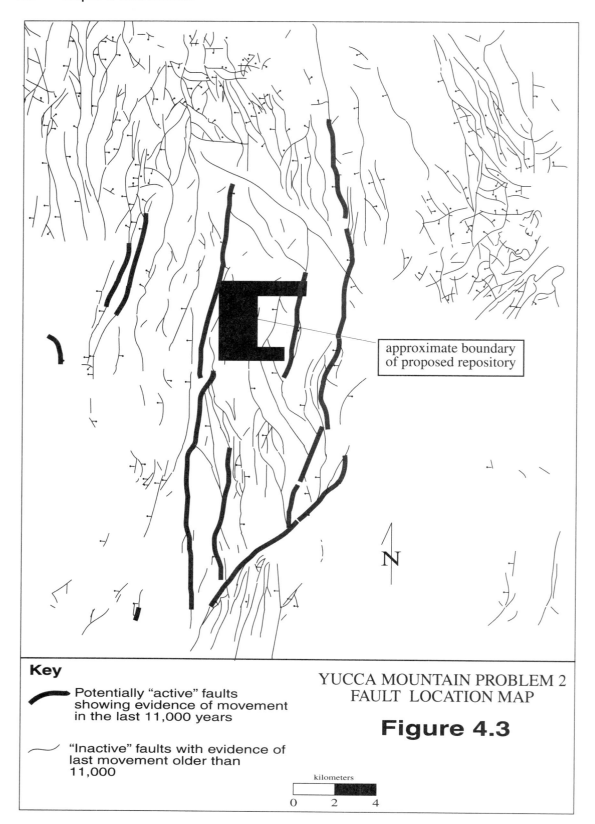

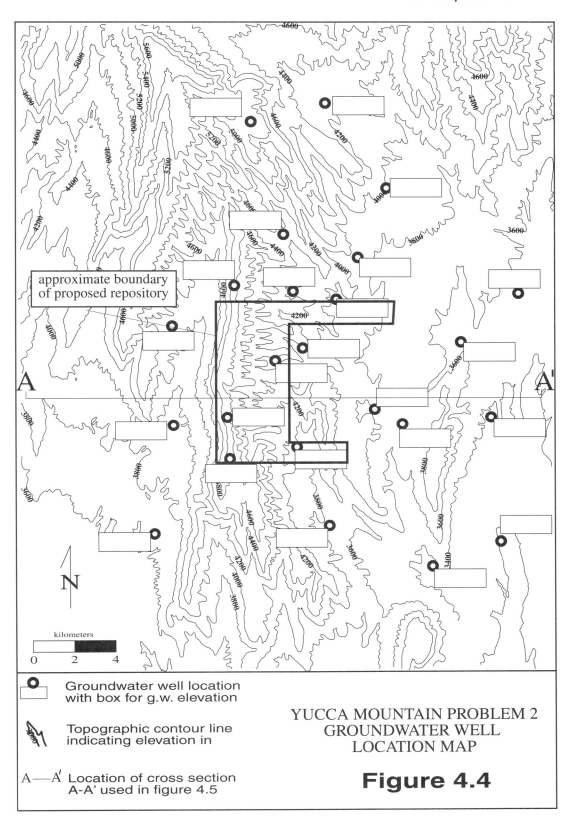

114   Chapter 4: Yucca Mountain

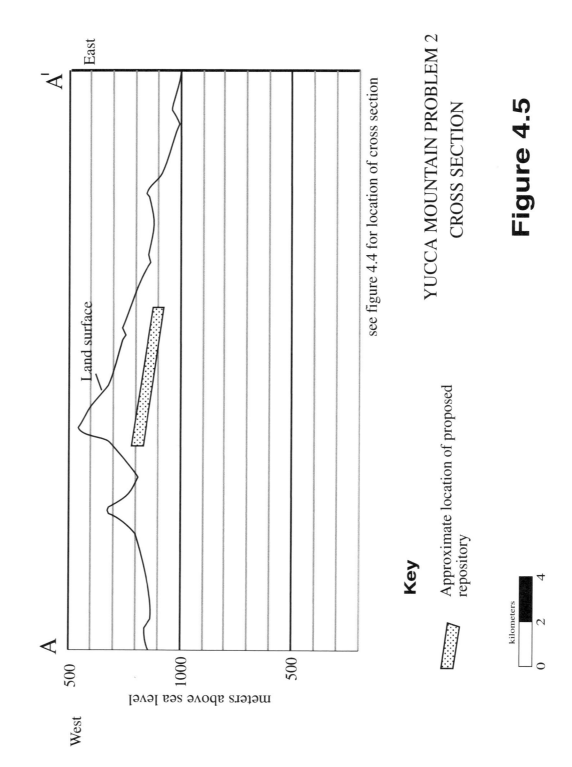

Figure 4.5

YUCCA MOUNTAIN PROBLEM 2
CROSS SECTION

Nuclear Waste Repository Research Problem 3: Transportation of Nuclear Wastes

Introduction and Advice

Before you start the research, examine **Study Notes 4.7** and **4.8** for some guidance about what you need to know before you start and what you should know when you are done. In **Study Note 4.9**, we provide a table you can use to keep track of assignments and their due dates.

> **Study Note 4.7: As you begin the research, you will need to have some prior knowledge in the following areas:**
>
> 1. A basic understanding of the complex problems associated with generation of nuclear waste.
> 2. Knowledge about the types of nuclear waste produced and the risks from radiation.
> 3. An introduction to the complex tradeoffs involved in determining how to deal with nuclear waste.

> **Study Note 4.8: What you should be able to do after completeing the research.**
>
> 1. Describe the plans for transportation of nuclear waste in the United States.
> 2. Develop a transportation plan for one of five areas in the United States.
> 3. Determine the hazards associated with the transportation plan and investigate alternatives.

> **Study Note 4.9:** Use the table provided below to list the research assignments given by your instructor. Make notes on this table about the types of work you will be expected to submit for a grade.

Research Assignments	Specific Instructions for Completing Assignment	Completion Date
Phase 1		
Phase 2		
Other Assignments		

Chapter 4: Yucca Mountain

What's the Problem?

So far, we have been studying the suitability of Yucca Mountain as a repository for high-level nuclear wastes. As we study the site itself, we also have to think about the plans for transportation of nuclear waste in the US. Remember, Yucca Mountain is in Nevada. In 1995, there were over 1,000 metric tons of radioactive waste stored at nuclear power plants in New Jersey. How would you move this high-level radioactive waste from New Jersey to Nevada? Your responsibility as a member of the scientific team is to decide if it even makes sense to transport the nuclear wastes from all of the current storage sites to Yucca Mountain in Nevada. If it does make sense, then you have to decide how to schedule such a massive move.

Not much to think about there, obviously, just outline some transportation plans. Then carefully think through the likely hazards associated with each plan. Well, we won't bore you with complexity just yet. Read the Background Information, and then work through the simulation. You can decide for yourself if it is feasible to transport nuclear wastes with an acceptable risk.

Background Information

The vast majority of the high-level nuclear waste in the United States comes from nuclear power plants, and it is this waste that is intended for the proposed nuclear waste repository. About 2000 metric tons of high-level nuclear waste is produced by the nuclear energy industry each year, and, at present, 32,000 metric tons of high-level nuclear waste is stored on-site at nuclear power plants around the United States. This waste, along with high-level nuclear waste from several nuclear power plants from Pacific nations, would be buried in the proposed nuclear waste repository at Yucca Mountain. Nuclear waste would be transported in large casks built to withstand significant impacts and heat. Transportation could occur either by truck or by rail. The Department of Energy (DOE) estimates that a repository would receive about 6,200 truck shipments and 9,400 rail cask shipments of spent nuclear power plant fuel, possibly along with a number of shipments of high-level radioactive weapons waste and "miscellaneous wastes requiring geologic disposal." The nuclear waste would, of necessity, pass through a number of cities, although the exact number and location of cities will depend on the routes and mode of transport, which have not yet been decided upon (State of Nevada Nuclear Waste Project Office, May 17, 1998).

There are advantages to both truck and rail transport for the waste. Many nuclear power plants do not have easy access to rail lines, so truck transport is much less expensive. At present, the Yucca Mountain area does not have rail access either. Rail shipments can be larger, meaning that fewer total shipments would be required. However, a decision would have to be made whether to build a rail line to Yucca Mountain or to transfer the waste to trucks for the final part of the transport, which is the current plan.

Many communities do not want the waste to pass through or near them, so the choice of routes will be a very politicized decision. There have been no traffic accidents leading to the release of nuclear waste into the environment since nuclear waste shipments first began in 1962. However, there would likely be more shipments in the first year or two of nuclear waste transport to Yucca Mountain than have occurred in the time since 1962. In addition, the transport distances would, on average, be much longer for the new shipments (State of Nevada Nuclear Waste Project Office, May 17, 1998).

Your job as a member of the research team is pretty simple. Just choose one of the five rail sectors or the "highway net" on the maps in the **Transportation room**, and determine a sequence and schedule for nuclear waste to be transported across it. We have already determined the "routes" – along superhighways and main rail lines. The goal is to get the waste to Yucca Mountain. As you look at the transportation map, you will notice some areas produce large amounts of nuclear waste, while others produce none. The Midwest, which produces little waste, is still en route for the nuclear waste coming from the East Coast.

A number of government studies concerning the risks of transportating nuclear waste have been made. There is a risk of accidents that might release large amounts of nuclear waste into the environment, and there is a risk of residents contracting cancer from the radiation given off by the waste as it passes through an area. The calculation of these risks is a very complicated process, so there is a range of calculated risks, depending on the parameters used in the calculations.

The risk of accidents is based largely on past experience of transportation of hazardous material. The risk is affected somewhat by which routes are chosen (and they have not yet been officially chosen) and by the regulations related to the transport. The risk related to the radiation itself is largely controlled by the type of packaging and the potential for gases to escape from it. In addition, there is always the small possibility of a terrorist attack on a shipment.

Based on data collected on nuclear waste shipments between 1971 and 1990, DOE calculated accident and incident rates for commercial spent fuel shipments to a repository. For truck shipments, they estimate 0.7 accidents and 10.5 incidents per million shipment miles, while, for rail shipments, they expect 9.7 accidents and 19.4 incidents per million shipment miles. There were not very many spent fuel shipments and accidents between 1971 and 1990, so DOE compared these accident/incident rates to those for large commercial trucks and general rail freight movements. DOE decided that accident rates for general truck and rail transportation should be used instead, in order to provide a more conservative estimate. They recommend the use of a truck accident rate of 0.7 - 3.0 accidents per million shipment miles and a rail accident rate of 11.9 accidents per million shipment miles (State of Nevada Nuclear Waste Project Office, May 17, 1998).

To figure out how many accidents might be expected, multiply the accident rate for the shipment mode by the number of shipment miles. So, if all spent fuel were to be shipped to the repository by truck in large-capacity casks (probably a worst-case scenario), about 46,000 shipments over 100 million shipment miles would occur, and we would expect between 70 and 300 accidents with over 1,000 incidents expected over the operating life of the repository.

So what are the possible contaminants that might be released? A ten-year-old spent fuel assembly can contain about 26,000 curies of strontium-90 (plus many thousands of curies of other dangerous isotopes). The strontium-90 and most of the other dangerous radionuclides are in solid fuel pellets and are not easily dispersed. A severe accident could cause a release of fuel mixed with smoke in a fire. These particles could then be inhaled or enter the soil and contaminate the food chain. Such an accident is unlikely, but the costs, both economic and human, would be severe, with a long period required for cleanup and determination of possible health effects on many people (State of Nevada Nuclear Waste Project Office, May 17, 1998).

Obviously, it is critical that the shipping containers be strong and capable of preventing such a release. The NRC requires that a cask be able to withstand, in succession, the following four tests: (1) a drop from 30 feet onto an unyielding surface; (2) a drop from 6 feet onto a spike (a puncture test); (3) a 30 minute fire at 1425 degrees (F); and (4) a 30 minute submersion in three feet of water. No design has yet been decided upon for shipments, but any design will have to meet these criteria (Nuclear Regulatory Commission, 1998).

There is a danger of radiation exposure for waste handlers, drivers, and the general public, even during routine (non-accident) conditions (see the book: **Radiation Effects**). To reduce this risk, shipping containers are shielded, because even after ten years of cooling of the waste, a person standing one yard away from an unshielded spent fuel assembly could receive a lethal dose of radiation (about 500 rems) in less than three minutes. A 30-second exposure (about 85 rems) at the same distance could significantly increase the risk of cancer and/or genetic damage. The surface dose rate of spent fuel is very high (~10,000 rems/hour), and shipping containers with enough shielding to completely contain all emissions would be too heavy to transport economically. Federal regulations allow shipping casks to emit 10 millirems/hour at 2 meters from the cask surface (Nuclear Regulatory Commission, 1998), equivalent to about one chest x-ray per hour of exposure. Such emission might become significant if a transport vehicle were stuck in a traffic jam with cars (and their drivers) all around it. There is some evidence that low levels of radiation can have long-term health effects. However, most of the public would not be in a situation to have repeated low-level exposures, so for most, the largest danger would be from a single low-level exposure during a traffic jam. The drivers and handlers would, of necessity, have to receive better shielding.

Given this information, think about what is an acceptable risk for the shipment of nuclear waste. Most people will accept a greater chance of something happening if the negative consequences of the occurrence are not too bad. If the negative consequences are extremely bad, most people prefer the chances of this occurring to be very low. There are many other reasons to alter one's perception of acceptable risks, such as whether or not one receives any benefit from the action (i.e. whether the nuclear waste leaves your area versus having it come into your area), or whether one has any personal experience with nuclear materials (e.g. whether one is a nuclear testing survivor or a nuclear physicist).

Before you start the research, review the processes that lead to the production of nuclear waste and the precautions that need to be taken when storing and transporting such wastes. You should also choose at least one transportation region.

Research Questions

We envision two broad areas of research for this problem. The first area is an analysis of sequencing and scheduling transportation. The second area is an estimation of hazards involved in the many steps of the transportation process. Thus, we see the research unfolding in two phases.

> **Phase 1**— Developing a schedule for transporting nuclear waste.
> **Phase 2**— Estimating transportation hazards.

Before you enter the simulation, take a look at these phases and their respective research activities.

Phases of Research

Phase 1 – Develop a schedule for transporting nuclear waste.

Enter the **Transportaion room**. You will choose either a rail sector of the United States (NW, SE, Central, NE, SE), or the entire Highway network as shown on the transportation map, and determine a schedule for transportation of nuclear waste to Yucca Mountain. There are some problems you have to consider, however. Review **Research Note 4.1** before you start.

Research Note 4.1: Some of the important constraints on transporting nuclear waste materials.

1. All waste in your selected area must be transported to Yucca Mountain within a time period of 10 years. The routes have been pre-selected based on the quality of the highway or rail system available near the reactor

2. All the waste you must transport is the amount that is currently present. Note that by the end of 10 years, this amount may in reality be doubled as nuclear power plants continue to produce the waste.

3. You won't have more than ten metric tons of high-level nuclear waste per train shipment, or more than 2.5 metric tons of high-level nuclear waste per truck shipment. The actual weight will be much higher because of the weight of the nuclear waste containers. The actual shipment size for each site is predetermined.

4. The waste should not stop anywhere along the predetermined paths. A waste shipment that is stopped increases the risk factor for the run. A total risk factor below 50 for the run is acceptable. We are using a generic "risk" value here, since there are so many problems involved with having nuclear waste "delayed" somewhere on its way to Nevada. What do you think these risks are?

Keep these constraints in mind as you develop your transportation plan. Your main goals are to avoid piling transport vehicles up in one spot (which would inherently increase risks) and to move the waste within 10 years.

Chapter 4: Yucca Mountain

Task 1:

Select your US rail or highway sector. The transportation network for that sector will be shown, and the location of each of the reactor sites. Next to (or above) each reactor is the site scheduler (a sliding red button on a green ruler). As you move the site scheduler button, the interval between shipments is displayed. Currently the selectors are all set for a 1 day interval. That is, a shipment should leave the reactor on a daily basis. However, there are a limited number of trucks and trains suitable to transport the waste, so they must also return to the reactor site before more shipments can be moved. Only loaded vehicles are shown in the simulation.

If you click on the start button, the simulation starts moving waste from each reactor towards the sector exits. The total number of shipments left at a site are shown above each sector. Trouble (increased risk) occurs if a shipment has to stop to allow another shipment access to an intersection or if a delay occurs for some other reason. You will see this happen if you observe an explosion symbol replace your shipment along the route. The risk increases if this happens.

You can also sequence the reactors by turning the shipments from them off or on. This could allow you to move small reactor amounts rather quickly, then focus on larger reactors later. To turn a reactor off or on, click on the green ruler at each reactor (it turns red).

You can reset the simulation with the RESET button if the risk gets above 50 before the waste is transported.

Try to limit risk in the simulation. Try to select a combination of intervals and sequences that will allow movement of the waste shipments out of the sector with a risk level below 50.

What strategies did you discover that helped determine an interval setting for a particular reactor?

_____
_____
_____
_____
_____
_____

Did sequencing work? Describe which areas were best suited for sequencing.

_____
_____
_____
_____
_____
_____

How did you decide to set intervals or sequences in crowded areas ?

Phase 2 — Estimating Transportation Hazards

Now let's calculate what the risks might be. You might want to review the **Background Information** and some of the electronic books related to radiation before starting work on the following research tasks.

Task 1:
Calculate the amount of exposure a person would receive if stuck 2 meters away from a transport cask of spent nuclear fuel for 2 hours. How much radiation would you receive if a nuclear waste truck broke down and spent the night (8 hours) parked 2 m away from your truck in a truck stop? Now, put this information in perspective by comparing it with normal background doses (see the book: **Radiation Effects**). What percentage increase would this be over normal background radiation? Would this dose exceed the legal limit for the normal public?

Task 2:
One study (Halstead and Ballard, 1997) suggests there is a danger of two radioactive releases per million miles of waste transport. Estimate how many radioactive releases you can expect, based on this figure. Why is the number of deaths from these releases very hard to quantify? In other words, what factors would determine whether a radiation release would cause deaths or injuries?

Chapter 4: Yucca Mountain

Task 3:
Building on information from the previous task, decide if this is an acceptable risk. What would be an acceptable risk?

Task 4:
Using the methods described above, calculate how many accidents would be expected under the expected shipment ratio of 88% by rail and 12% by truck. Figure on a total of 100 million shipment miles.

Task 5:
In its research, the US Department of Energy has calculated that it will take 6,217 truck plus 9,421 train shipments to move all the nuclear waste presently stored on site at nuclear power plants to Yucca Mountain. Does this change your answer? If so, how and why?

Task 6:
Some people have suggested the US should not move the waste from the on-site storage at nuclear reactors, at least at present. They suggest that the waste will become less radioactive with time, and we would be better off waiting 50-100 years for some of these short-lived radionuclides to decay.

Describe factors that would affect the risk if the waste were simply left in place at the nuclear power plants.

Task 7:

Predict the factors that might make it safer to move the waste 100 years from now. Also, describe factors that might make the on-site storage of nuclear waste more dangerous than transporting it to a nuclear waste repository over the next 10 years.

References

Halstead, RJ, and Ballard, JD, 1997, Nuclear Waste Transportation Security And Safety Issues: The Risk of Terrorism and Sabotage Against Repository Shipments, Expanded Version of Presentation at 1996 Southwest Counter-Terrorism Training Symposium, Las Vegas, Nevada, September 24, 1996; http://www.state.nv.us/nucwaste/trans/risk01.htm

Nuclear Regulatory Commission, U.S., 1998, Standard Review Plan for Transportation Packages for Spent Nuclear Fuel, Nuclear Regulatory Commission NUREG-1617, Draft Report for Comment, Washington, D.C., also available at http://www.nrc.gov/NRC/NUREGS/SR1617/sr1617.html.

State of Nevada Nuclear Waste Project Office, May 17, 1998, Fact sheet: Transportation of spent nuclear fuel and high-level radioactive waste to a repository,
http://www.state.nv.us/nucwaste/yucca/trfact01.htm.

In the Know about Transporting Nuclear Wastes

Suppose you have to report to a committee that is composed of members of the United States House of Representatives and the Senate. They are not technical experts, but their staff members will require a technical report which they will have reviewed by their expert consultants. You have to write this technical report, as well as a less technical set of responses that will answer likely questions from the congressional panel. Organize your technical report and nontechnical responses around the following questions.

1. What are the major problems associated with nuclear waste shipments?

2. Why are these problems complex?

3. What are the problems involved with scheduling waste shipments from a sector?

4. What are the possible hazards from nuclear waste shipment?

5. How do you justify your estimations of the risk?

6. How have you determined the complex tradeoffs involved in dealing with existing and future nuclear waste?

7. What plans have you developed as alternatives to immediate shipments from current storage sites to the nuclear waste repository?

Nuclear Waste Repository Research Problem 4: Risk of Volcanic Activity

Introduction and Advice

Before you start the research, examine **Study Notes 4.10** and **4.11** for some guidance about what you need to know before you start and what you should know when you are done. In **Study Note 4.12**, we provide a table you can use to keep track of assignments and their due dates.

Study Note 4.10: As you begin the research, you will need to have some prior knowledge in the following areas:

1. A basic understanding of volcanism
2. Some knowledge of methods for isotopic dating of rocks.
3. An understanding of the possible effects of an eruption upon a nuclear waste repository.

Study Note 4.11: What you should be able to do after completing the research.

1. Explore the locations of volcanoes in the Yucca Mountain area and sample volcanoes for dating.
2. Analyze data on ages of volcanic rock and calculate dates and analytical errors for eruptions.
3. Use data on locations of volcanoes and dates of volcanic rock to determine the likelihood of an eruption at Yucca Mountain.

> **Study Note 4.12:** Use the table provided below to list the research assignments given by your instructor. Make notes on this table about the types of work you will be expected to submit for a grade.

Research Assignments	Specific Instructions for Completing Assignment	Completion Date
Phase 1		
Phase 2		
Phase 3		
Other Assignments		

What's the Problem?

If you have worked through Yucca Mountain **Research Problems 1-3**, you are undoubtedly getting a sense of the complexity of making a decision about the viability of choosing the Yucca Mountain site as the high-level nuclear waste repository. Let's make life even a little more complicated and study the potential for volcanic eruption at or near the site.

We have done some preliminary work on volcanic activity in the earlier **Research Problems**. Now, we need to imagine a catastrophic scenario and estimate the risk that this scenario could happen. You will probably agree a volcanic eruption near the repository has the potential for destroying the integrity of the storage areas and releasing radioactive wastes that could be dispersed widely by ash and small particulates carried into the atmosphere. As they fall back to earth, they would be a very real radioactive *fallout* that posed a variety of dangers to humans and ecosystem structure. Could this scenario occur? Interestingly, another possibility is that a small eruption could fill the repository with lava and seal the radioactive wastes tighter than we ever could.

Well, one strategy for determining the likelihood of volcanic eruptions is to examine the dates of past eruptions and see if there are trends indicating a pattern of volcanic activity. For example, you could explore the locations of volcanoes in the Yucca Mountain area, sample rocks from these volcanoes, and date the rocks. If you were to find that volcanic eruptions are getting more recent as you move towards Yucca Mountain, you might worry about increased probability of an eruption occurring at the site. On the other hand, if the eruptions were more recent as you moved away from Yucca Mountain, you might conclude that the pattern of eruptions is moving away from the site and the chances of an eruption at the site are not significant.

All you need is a *guide to dating*. Rocks, that is! We have oversimplified a little in this first section. You will find that quite a bit is involved in looking for volcanic lineaments, dating volcanic rocks along the lineaments, and using the dates to estimate the likelihood of eruption along the lineaments. However, this sequence of activities does make sense as a way to phase research activities. Start by reading the **Background Information**, and then plunge into the simulation and figure out the threat of volcanic eruption at or near the Yucca Mountain repository.

Background Information

The determination of the volcanic risks to the proposed Yucca Mountain high-level nuclear waste repository requires finding out where and when volcanism might occur, as well as what sort of eruption this might be. In this laboratory exercise, you will investigate the volcanic hazards at Yucca Mountain. In order to do this, you will determine whether there are any volcanic lineaments in the area, and then you will date the eruptions to see if there is any general age progression of the eruptions in a particular direction. Finally, you will evaluate this information in terms of what the likely style of eruption might be in order to evaluate the hazards at the Yucca Mountain site.

There are several ways to approach the problem of dating a volcanic event. One way is to simply look at the shape of the vent area. Is it deeply eroded, or un-eroded and fresh-looking? Are the flanks of the volcano covered by sediments? These would give some idea of the relative age of the vent. This technique is used by volcanologists to get a general idea of the ages of volcanoes. It requires little to no instrumentation, but it provides only a qualitative answer - the volcano is "pretty old" or "pretty young."

Another, somewhat more quantitative, way to determine the age of a volcano is through a technique called secular variation which uses a branch of geology called paleomagnetism. The Earth's magnetic poles wander over time, moving by a few degrees latitude and longitude from their present locations. Occasionally (at irregular intervals on the order of millions of years), the north and south magnetic poles actually trade places. Lavas contain a significant amount of iron, and when they cool the iron acts somewhat like a compass and locks in a record of the orientation of the magnetic poles at the time the lava cooled. Scientists have developed a record of the places that the magnetic poles have moved over time (called a secular variation curve) by analyzing samples of lavas which they have dated by independent means. They can then drill oriented samples from a volcanic rock, analyze the magnetic directions carried by that rock, and compare them to the secular variation curve. If they have a good record, they can determine the age of the volcanic event in this way.

Scientists have used both visual inspection and secular variation methods at Yucca Mountain. Neither has been very satisfactory. The visual inspection technique yields relative ages, but erosion of volcanoes is very slow in the arid environment around Yucca Mountain, so it is difficult to calibrate this relative dating technique. The secular variation technique's utility has been hampered by incomplete records. There are several different times in the past when the magnetic poles have been in particular locations, and without a continuous record, the isolated samples yield several possible, non-unique dates.

Since these two techniques have not provided satisfactory results, geologists have settled on a third, quantitative technique called isotopic dating. This technique, described in the book **Isotopic Dating**, uses the rates of radioactive decay of particular elements to date the rocks. It provides quantitative answers, but also has some difficulties that make the dates a bit harder to interpret. In this exercise, we will use K/Ar dating to determine the dates of the eruptions of the basaltic volcanism in the Yucca Mountain area.

As you begin this research problem, make sure you have a basic knowledge of potential volcanic lineaments in the Yucca Mountain area. It will also be critical for you to have some knowledge of how radioactivity occurs. Finally, it will be valuable to have a general familiarity with the Yucca Mountain Project. For background information or a review, please refer to earlier **Research Problems** on Yucca Mountain as well as to the electronic books in the *Bookshelf*.

Research Questions

In Yucca Mountain **Research Problems 1-3**, we offered an overarching research focus that could be studied by answering specific research questions. The research questions were always tied to phases of research. In the What's the Problem section above, we suggested three phases of research activities that had to be completed in order for us to get at the overarching concern: "What is the likelihood of volcanic eruption in the vicinity of Yucca Mountain?" You recall that these phases were:

- **Phase 1** — Looking for lineaments.
- **Phase 2** — Dating the volcanic rocks from volcanoes in the lineaments.
- **Phase 3** — Using the dates of rocks to estimate the likelihood of volcanic eruption.

Before you start these phases of work, figure out the logic of the sequence of phases and how data from one phase is used in the next phase. Make a list of question that come up as you think about each phase of research activities. You can use the table in **Research Note 4.2** to write some of these down. Then, as you proceed into the phases of research, decide if your logic and questions provided a good framework for thinking about the problem being studied.

Research Note 4.2: Research logic for your study.

Phases	Logic for Phase and Types of Data Gathered	Questions that Need to be Answered
Lineaments		
Rock Dates		
Likelihood of Eruptions		

Phases of Research

Phase 1 — Looking for Lineaments.

First, read the information on volcanoes in the Yucca Mountain region in the **Volcano Primer** book. Knowing that basaltic eruptions are what has been occurring for the past 10 million years, look on the geologic maps to determine where these eruptions have occurred. Read (or re-read) the discussion of possible volcanic lineaments in the Yucca Mountain area, and look for possible lineaments yourself, using the geologic maps. If you completed **Research Problem 2** for Yucca Mountain, you have already done this first part. Refresh your memory by answering the following questions. If you did not do **Research Problem 2**, this portion will require you to do a little work to get ready for the latter parts of this exercise.

Task 1:
On the attached map (**Figure 4.6**), color in the locations of basaltic eruptions. Indicate any possible lineaments that you see. Describe these lineaments. Are they distinct, or are they rather fuzzy, so you are not sure if they are a lineament? Are any of them close to Yucca Mountain?

_____
_____
_____
_____

130   Chapter 4: Yucca Mountain

Task 2:
Describe what you believe controls the orientation of these lineaments. Look at the data in the **Map** room to see if there are geologic structures that might control them. What are these structures?

_____

_____

_____

Phase 2 — Dating the Volcanic Rocks.

Now, you have identified some possible lineaments. The next question is whether the volcanism along these lineaments has any age progression. If the locus of volcanism is moving away from the Yucca Mountain area, the danger to the proposed repository would be less than if it is moving toward the site. To answer this question, you must determine the dates of the eruptions at each of the centers.

The K/Ar dating technique uses a few equations described in **Research Note 4.3**. Please consult this note before you proceed with the dating study. We think it also would be a good idea for you to read the **Isotopic Dating** book in order to more fully understand the technique.

Research Note 4.3: Equations used for the K/Ar dating technique.

The equation for calculating the age of a rock using the K/Ar isotopic system is the following:

$$t = \frac{1}{\lambda} \ln\left[\frac{^{40}Ar^*}{^{40}K}\left(\frac{\lambda}{\lambda_e}\right) + 1\right]$$

where t is in years, λ is the total decay constant for 40K (5.543×10^{-10} y^{-1}), and λ_e is the decay constant for ^{40}K going to ^{40}Ar (0.581×10^{-10} y^{-1}).

To calculate $\frac{^{40}Ar^*}{^{40}K}$, use the following equation:

$$\frac{^{40}Ar^*}{^{40}K} = \frac{mol\ ^{40}Ar^*/g}{\%K \times 0.001167 \times 39.9623 A}$$

where mol $^{40}Ar^*$ is a reported value from the isotopic analysis for the number of moles of radiogenic argon, %K is calculated from the reported %K_2O, 39.9623 is the atomic weight of potassium, 0.0001167 is the abundance of ^{40}K expressed as a decimal fraction, A is Avogadro's number (6.022×10^{23} molecules/mole), and the factor 10^4 changes the percent concentration of K to parts per million, to match the units of ^{40}Ar.

The percent of K is calculated by the following equation:

$$\%K = \frac{2(atomic\ mass\ K)}{atomic\ mass\ K_2O} \times \%K_2O$$

where atomic weight of K = 39.9623 amu and atomic weight of O = 15.9994 amu, and the %K_2O is reported.

Task 1:
You will use all of the equations in **Research Note 4.3** to calculate the ages of the volcanic rocks of the Yucca Mountain area. Look again at your colored geologic map of the Yucca Mountain area. In order to determine the age of a volcano, scientists typically sample several different places on the volcano, date them, and then statistically weight the mean age. You will need to sample sites on each volcano, and will virtually send them off for analysis. You will receive the data back, and then calculate dates and statistical errors for the rocks.

Go to the Isotopic Dating map of the volcanic centers of the Yucca Mountain area in the map room. Navigate through the virtual environment to each volcano. At each volcano, you will sample two locations. Keep track of the sample numbers associated with each volcano in **Table 4.2** below, as the data will be reported back to you simply by sample number. Sample all of the volcanoes in the area. When you are done and exit the simulation, the samples will automatically be sent off for analysis, and the data reported back to you. Write your data down on the table below.

Table 4.2: K/Ar Dating Data

Sample #	Location	%K_2O	mol ^{40}Ar/g	% ^{40}Ar

Task 2:
Using the data just collected, calculate the dates for each volcanic center, and write them in **Table 4.3**. Look at the dates for each volcanic center. You will notice that they are not always in perfect agreement with each other. This may seem counter-intuitive, as a single eruption (which is what is thought to have formed most of these vents) should produce lavas that all have the same age. It turns out that no analysis is perfect. There is always a certain amount of analytical uncertainty due to instrument dynamics and electrical current fluctuations. In addition, there can be some geological effects, such as minor alteration of the rock, or inherited Ar from a slight amount of melting of an older crustal rock during ascent of the magma.

Therefore, in order to evaluate the ages, one must look at the estimated errors in the sample. In order to evaluate this, one can simply evaluate the amount of fluctuation of the measuring instrument when it is measuring a standard of known age. This will give a good idea of the total instrumental variance. It is much more difficult to evaluate the geologic sources of variance. One does this by looking at the analyzed rock under a microscope in order to find any possible alteration of the rock, and even by chemically analyzing the rock's crystals to see if they show signs of alteration. The problem of excess argon coming from other crustal rocks is very difficult to evaluate by the K/Ar method. Another method, called the 40Ar/39Ar method which uses radiation to turn some potassium into

39Ar, is commonly able to detect this excess argon. Some $^{40}Ar/^{39}Ar$ dates have been made on the rocks you dated, and they generally indicate there is no excess argon. Only Lathrop Wells cone, the youngest volcano in the area, has possible problems with this.

For our purposes in this laboratory, we will simply estimate the internal errors, which are those caused by the analytical process. To do this, we will use a regression equation that relates the estimated standard deviation (in percent) to the % $^{40}Ar^*$. **Research Note 4.4** describes the method.

Research Note 4.4: The regression equation that relates the estimated standard deviation (in percent) to the % $^{40}Ar^*$.

The equation is derived by doing repeated analyses of known standards of various ages on the instrument used to analyze your samples. Then, the many ages obtained for each standard are plotted on a histogram. They form a Gaussian distribution (like a 'bell curve'), and the range in which 68% of the ages lie (called the 'one sigma' standard deviation) is determined. This range is then plotted against the % 40Ar* values. This is done because the primary predictable variable in the analytical error is how much 40Ar is available to be measured. The other errors, such as instrument drift, are random, and are taken into account by the scatter of data on this plot. The slope of the best fit regression line through the points on this graph provides an equation that is then used to estimate the standard deviation for other samples. This equation, for samples of concern to this lab, is **log y = 1.613 - (0.839 × log x),** where y is the estimated standard deviation in percent and x is % 40Ar*. The correlation coefficient for this line is 0.909, where 1 would mean that all data points lie directly on the line. This is a pretty good correlation coefficient, and indicates that the data fit pretty close to the line. The small amount of scatter is due to factors such as instrument drift and fluctuations in the electrical current during the analysis. Calculate the estimated standard deviations for your calculated dates using this equation, and write them in **Table 4.3**. Then, multiply this % standard deviation by the calculated ages to get the estimated standard deviation in years and write it in **Table 4.3**. This is sometimes called the "plus or minus" on the date, meaning that 68% of the analyses would lie within this range above or below the date determined. Another way to say this is that there is a 68% confidence that the actual date is within this range. A "two sigma" (2σ) error for a Gaussian distribution would be twice this range. This is sometimes used for dates that require very high precision.

Table 4.3: Calculated K/Ar Dates and Errors

Sample #	Location	Calculated Date	Estimated standard deviation (%)	Estimated standard deviation (in years)	Best Estimated Age

Task 3:

Now, compare the different dates you have for each volcanic center. Do they fall within analytical uncertainty of each other? In other words, if you give them an age range equal to your calculated standard deviation around each age, do the age ranges overlap? If not, explain what factors might cause this disparity.

_____
_____
_____
_____

Task 4:

Come up with a best estimated age for each volcano. To do this, use the following weighted mean values equation:

$$\text{best estimate date} = \frac{\frac{1}{SD_1}}{\frac{1}{SD_1} + \frac{1}{SD_2}} date_1 + \frac{\frac{1}{SD_2}}{\frac{1}{SD_1} + \frac{1}{SD_2}} date_2$$

where SD_1 and $date_1$ are the estimated standard deviation (in years) and calculated date for sample 1, respectively, and SD_2 and $date_2$ are the same for sample 2. This equation gives you a date for the volcano that is weighted according to the quality of the analyses. Write the best estimate date for each volcano in **Table 4.3**.

Chapter 4: Yucca Mountain

Phase 3 — Using the dates of rocks to estimate likelihood of volcanic eruption.

Phase 3 focuses on how we determine the likelihood of a volcanic eruption near the proposed Yucca Mountain repository.

Task 1:
Go back to your map of the volcanic centers of the Yucca Mountain area (**Figure 4.6**). Plot the dates you obtained for each center on the map. Do you see an age progression for the volcanism in the area? If so, what is it? If not, is the pattern of ages entirely random, or do you see some order (such as that certain lineaments are younger and others older)?

_____
_____
_____
_____
_____
_____

Task 2:
If you see an age progression in your findings from **Task 1**, this may indicate volcanism is more likely to occur in the direction of younger volcanoes, and less likely in the other direction. If you do not see an age progression, then you must treat the volcanism as equally likely along any portion of any observed lineaments. Read the book: **Volcanic Risks**, and think about the significance of your findings.

Based on your dates and observations of possible lineaments, is volcanism more or less likely to occur in the proposed repository site than elsewhere in the Yucca Mountain area?

_____
_____
_____

Task 3:
Finally, let's make a rough estimate of the volcanic risk in the Yucca Mountain area. There have been 9 volcanoes that formed in the past 5 million years within the Crater Flat area. The area of volcanism is a NE elongate rectangle roughly 14 km wide and 20 km long, or 280 km2.

First, calculate the recurrence interval for volcanism for the Crater Flat area (includes Yucca Mountain), which is

$$\text{area recurrence interval (events / year)} = \frac{\text{\# of eruptions}}{\text{Total time period}}.$$

For our purposes, assume that each volcano only erupted once. This may or may not be true — scientists debate this point for Lathrop Wells cone.

a) Determine the recurrence interval for volcanism in the Crater Flat area.

b) Next, calculate the likelihood of an eruption in any particular location. Assume that an eruption would devastate a one kilometer area (and cause the release of radioactive waste). Thus, the equation for this is:

$$local\ recurrence\ interval\ (events\ /\ year) = \frac{1\ km^2}{(280\ Km^2) \times (area\ recurrence\ interval)}.$$

What is the likelihood of an eruption within any one square kilometer area in the Crater Flat area?

_____
_____
_____

c) This value is a first-order indication of the risk at Yucca Mountain if the volcanism is truly random. If volcanism is concentrated along lineaments, the risk is greater along these lineaments, and lower away from these lineaments. Evaluate the volcanic risk to the proposed high-level nuclear waste repository at Yucca Mountain in these terms.

References

The K/Ar data used in this laboratory are derived from:
Fleck, RJ, Turrin, BD, Sawyer, DA, Warren, RG, Champion, DE, Hudson, MR, and Minor, SA, 1996, Age and character of basaltic rocks of the Yucca Mountain region, southern Nevada, Journal of Geophysical Research, v. 101, p. 8205-8227.

In the Know about Risk of Volcanic Activity

The scientific team has members who are experienced volcanologists as well as other researchers who know very little about how to estimate risk of volcanic activity. You are asked to make a presentation to the whole research team. In this presentation, you have to justify the use of volcanic lineaments and rock dating as the best method for collecting data that can be used to evaluate risk of volcanic activity near the repository. Since the audience is diverse in their expertise, use the following questions as a guide for developing your presentation. Basically, this means that you will have to answer each of the following questions in such a way that all members of the research team are convinced your approach was the best one.

1. What is the theoretical basis for isotopic dating methods?

2. How can you use isotopic dating methods to evaluate the age of volcanoes?

3. What are the other methods used for dating volcanoes, and why weren't these methods used?

4. How can you use the age of volcanoes to evaluate volcanic hazards?

5. What kind of structural controls might have affected the eruptions in the volcanic fields? What kinds of eruptions are most likely to occur in the Yucca Mountain area?

6. What are the possible effects of an eruption near the nuclear waste repository?

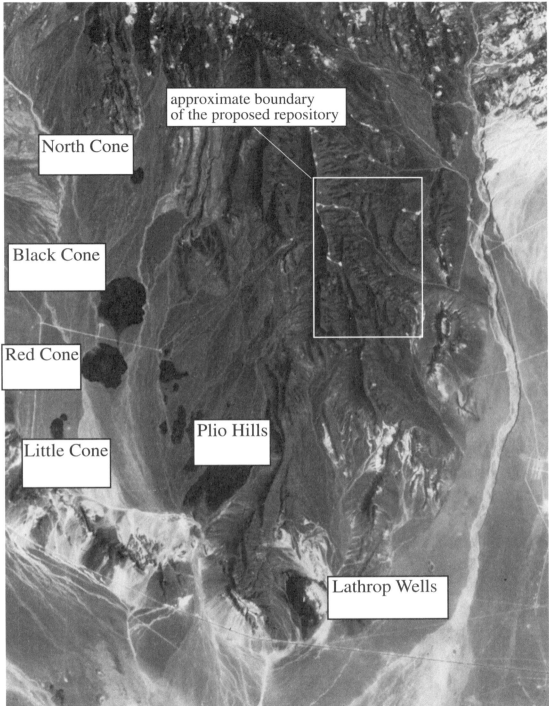

YUCCA MOUNTAIN PROBLEM 4
VOLCANO LOCATION MAP

Figure 4.6

SO, WHAT'S THE POINT?

As you think about the suitability of the Yucca Mountain site as a repository for nuclear waste, you might be interested to know there was a report of a tornado hitting a nuclear power plant in Ohio. A variety of newspapers reported that on June 24, 1998, the Davis-Besse Nuclear Station near Oak Harbor, Ohio, took a direct hit by a tornado. It appears the 900 megawatt reactor shut down automatically. No radiation leaks were reported.

How would you find out more about what happened to the Davis-Besse Nuclear Station? For example, one newspaper reported that the tornado caused a number of equipment problems and some real concern among the operators. What sources of information would you use to assess the actual damage to the power plant? What data are necessary for you to predict tornado risks to both power plants and the spent nuclear fuel stored on the site? Use the following guidelines to begin your evaluation of the risk from major weather events to the current storage of nuclear fuels, to transportation of spent fuels to the Yucca Mountain Repository, and to the stability of the Yucca Mountain site.

- What sources of data would you examine in order to estimate and evaluate the impacts of major weather events on nuclear power plants and current storage facilities for spent nuclear fuels?
- How would you evaluate the data and information from various sources? For example, would you base any conclusions on only newspaper articles? What criteria do you use to evaluate information?
- What are the weather threats to transportation of nuclear wastes? How can transportation of nuclear wastes be implemented to minimize risks imposed by major weather events?
- See if you can find specific data on the Davis-Besse Nuclear Station on Lake Erie, particularly in the aftermath of the tornado on June 24, 1998. Using these data, try to solve the problems in the following hypothetical scenarios:
  1. Suppose the spent reactor fuel rods are stored in a pool that keeps them cool, and cooling is provided by electrical powered cooling pumps. A major storm causes damage that causes the power plant to lose electricity for critical systems. Furthermore, at least some of the systems are damaged by the storm. How many systems are in place to prevent the spent fuel pool temperature from rising to the point at which evaporation begins and there is a greatly increased risk for the spread of radioactivity?
  2. Nuclear power plants generally have emergency electrical generators that run on fossil fuel stored at the site. A major storm disrupts off-site power, turning on the emergency generators. However, the storm has damaged some of the components in the emergency generator system, and they are functioning below capacity. Are there other back-up systems? Sketch out the main systems and back-up systems of a nuclear power plant, and try to estimate the acceptable risks for sequential failure of the linked systems required to keep the plant from a catastrophic accident.
- Maybe you watch Real TV, the Discovery Channel, or the Learning Channel. Sometimes, these channels have shows about the wild and powerful side of nature in tornadoes, hurricanes, typhoons, and major shifts in weather caused by events like El Nino. Are major weather patterns predictable? If not, how can you calculate risks of impacts from storm systems on storage and transportation of nuclear wastes?

Mire to Fire: Coal Power

WHO CARES ABOUT COAL POWER PLANTS?

Do you have a light on while you're reading this? How about a computer, or the stereo? Maybe, a radio? Fixing popcorn in the microwave? How many electrical appliances are running while you read this paragraph?

Our point is that Americans use electricity in many, many facets of their daily lives. We want you to think for a moment about the origin of the electricity you are using right now. Sure, you can probably name the power company that sends out monthly bills to you, or whomever is paying for your electricity, but where does the company's electricity come from? If you have completed the Yucca Mountain simulation, you will remember how we asked similar questions about suppliers of your electrical power.

In the United States 50-60% of the total electricity is generated by power plants using coal. Estimates vary a bit, and you might find it interesting to check the percentage we report against several sources. You might also find it interesting to look at the percentage of electricity generated by coal-fired power plants in other countries. Going back to our original question about the origins of the electricity you use, is it likely that some of your electricity comes from power plants that burn coal?

We want you to do some thinking about how to study coal formations, the use of coal to generate power, by-products of burning coal, and potential impacts on humans and the environment. Search in the *Bookshelf* for sources of information. Also, consult the text or readings that have been assigned for the course in which you are using **VRX**. Below, we offer some questions that are meant to provoke thought from you as a citizen in a technologically-oriented democracy.

- What are the different kinds of coal, and what kinds are used most often in power plants?
- How is electricity generated by coal-fired power plants? How efficient is electricity generation by coal power plants in the United States? In other countries?
- What are the by-products of using coal to generate electricity? For example, how is coal mined, transported, cleaned, stored, burned, and what impacts and waste materials are created in these processes?
- How dangerous are the wastes from burning coal to produce electricity? Will the dangers decrease through time—for example, are countries likely to increase their use of coal for power generation? How can any dangers to humans and the environment be minimized?
- You discover there is a coal-fired power plant within 200 miles of where you live. What by-products of burning coal are likely to influence your quality of living? Be sure to think of both the positives and the negatives. For example, do coal power plants produce electricity at lower or higher costs than other fuels? Are the environmental impacts (including mining, processing, storing, burning, and waste removal) higher or lower than other means of generating electricity?
- Even if we stopped using coal to generate electricity (which is unlikely in the immediate future), will there be long-lasting environmental impacts related to our history of coal use and the current and projected use of coal in other countries?
- You are an advocate for sound environmental management and minimization of electrical power generation that creates dangerous by-products. If alternative methods of clean power generation are available, but will be expensive to develop and implement widely, how will you help inform the American population that their electrical energy costs are going to increase?

- After all is said and done, what level of risk will you accept? For example, would you accept the risk of increased greenhouse gases from the use of coal to generate power, the risk to others (like coal miners), various risks of environmental damage related to processing and burning coal, and health risks to you and your family?

WHAT DO YOU SEE IN THE REAL WORLD?

Coal is a fossil fuel, because it is essentially the remains of ancient plants. There were two major coal-forming periods during which the coal beds having the dominant economic value were formed. The Carboniferous Period in the Paleozoic Era lasted from 360-286 million years Before Present (BP). The Cretaceous Period in the last part of the Mesozoic Era occurred from 144-66 million years BP. During these periods, there were abundant swamps, bogs, and mires in which plants died and their organic remains were covered with water. Such material would have decomposed, but the amounts of organic detritus were sufficient that bacterial decomposition used up available oxygen and the decomposition process slowed or stopped. This happens when organic materials accumulate underwater faster than they can decompose.

Through time, the partially decomposed plant materials became an organic muck that can be covered and buried by sediments. Peat bogs are good examples of plant materials that accumulated in a water filled basin where decomposition was inhibited by lack of oxygen. As the water basin became filled with sediment, the plant materials became what we call peat.

However, in some locations, the plant materials became more deeply buried and compressed by layers of sedimentary rock that formed above the organic materials. Coal was formed by geological processes that compressed and heated the organic material, causing loss of water content and other volatile components like gases (oxygen, nitrogen, hydrogen). Different combinations of compression and heat led to different kinds of coal. Again, we encourage to you look up the kinds of coal, and note the kinds of compressive and thermal conditions that formed them. Take a look in the *Bookshelf* for resources related to coal and coal power plants.

Coal beds, ranging in size from enormous to relatively small, are the end result of coal-forming processes that occurred in the Carboniferous and Cretaceous Periods. If you look for the economically viable coal beds, you will find that most are in the northern hemisphere, with a large percentage of the known and probable undiscovered deposits falling within the borders of the United States, the Peoples Republic of China, and the area covered by countries which comprised the former Soviet Union. The world has more coal than any other fossil fuel, but it is not distributed evenly around the globe. This unequal distribution has implications for economic growth in different countries. Coal is also the dirtiest fossil fuel in terms of the amounts and relative environmental impacts of by-products from coal combustion. You might want to look up data on coal use in China, for example, and estimate the production of CO_2 from the extensive use of coal for energy production in that country. How does the level of China's CO_2 production compare to that of other industrialized nations? What are the implications for global agreements about control of greenhouse gases?

The processes of coal formation, the types of coal formed, and the use of coal for production of electrical energy are important to your life. There are direct effects, like the impact on our electricity use, and there are indirect effects, such as impacts of coal use on global economic interactions and production of greenhouse gases. We want you to have a chance to study coal formation and use in

142   Chapter 5: Mire to Fire

a coal-fired power plant. Unfortunately, the formation of coal we use has already taken place. It's very tough to travel back to the Carboniferous or Cretaceous Periods. Even if we could, our somewhat limited lifespans would not give us enough time to follow even a small segment of the process to completion. Try sitting around and watching a pond become a peat bog!

Instead, we offer you a virtual world, where time is not a particularly limiting constraint. We give you a Late Cretaceous mire. Now you can study coal formation from the start, instead of taking a modern coal bed and trying to work backwards to the environment in which the coal started its long transformation from plant materials. The research we offer in the virtual Cretaceous mire requires you to conduct a kind of "thinking" experiment. The mire provides a "what if" environment that you can explore. A key issue is that you need to be as rigorous and substantive in your research as you would be if you were studying a modern environment.

Once you've mucked about in the mire, you can move forward in time and discover where the coal exists now, and how the environmental conditions in the Cretaceous shaped the composition of the coal. Finally, you can put the coal into a furnace and try to generate some electricity. You may find managing stable electricity production is a difficult procedure, particularly during peak hours of electricity use or other times of high electricity demand (such as on hot days when air-conditioners are widely used), resulting in enormous electricity demands on power plants.

DEALING WITH COMPLEXITY

In this set of virtual worlds, you will be able to travel back in time to an environment which has the characteristics of producing organic sediments that can be transformed into coal. The first two simultions will focus on exploration of the late Cretaceous coastal wetlands environment. The third simulation takes you back to present time and allows you to study geological formations and land use in order to choose a site for a coal-fired power plant that has to be located near coal deposits. The fourth simulation setting is a coal-fired power plant. You have to control five subsystems to operate the plant in order to maintain stable power output: (1) coal storage and preparation, (2) the boiler-turbine subsystem, (3) water coolant subsystem, (4) the fly ash vacuum removal subsystem, and (5) the stack scrubber subsystem.

In these research settings, you will have to assume a number of roles. Your principal responsibilities will be to study the process of coal formation, and use your understanding of this process to think through the issues related to locating and operating a coal-fired power plant. This chapter provides the following research opportunities.

- **Coal Power Research Problem 1:** The Core of the Matter – Working in a Late Cretaceous Mire . . . 143
- **Coal Power Research Problem 2:** Predicting Coal Quality 155
- **Coal Power Research Problem 3:** Locating a Coal-Fired Power Plant 163
- **Coal Power Research Problem 4:** Operating a Coal-Fired Power Plant 180

Coal Power Research Problem 1: The Core of the Matter – Working in a Late Cretaceous Mire

Introduction and Advice

Before you start the research, examine **Study Notes 5.1** and **5.2** for some guidance about what you need to know before you start and what you should know when you are done. In **Study Note 5.3**, we provide a table you can use to keep track of assignments and their due dates.

Study Note 5.1: As you begin the research, you will need to have some prior knowledge in the following areas:

1. A basic understanding of how coal is formed.
2. Knowledge about the Late Cretaceous and coastal wetlands environments.
3. Some elementary knowledge about how to collect core data (stratigraphy and lithology).
4. Knowledge about how and why to construct a geologic cross-section.
5. An understanding of how to relate quality of ancient peat beds to quality of coal formed from these beds.

Study Note 5.2: What you should be able to do after completing the research.

1. Study landforms and vegetation patterns in order to determine likely areas of sedimentation and sources of materials for the sediments.
2. Access subsurface data by means of coring.
3. Analyze the compositions of cores and interpret sedimentation patterns.
4. Construct geologic cross-sections.
5. Interpret variation in peat composition in terms of environmental variations, specifically detrital origin of ash and marine influence on sulfur content.

> **Study Note 5.3:** Use the table provided below to list the research assignments given by your instructor. Make notes on this table about the types of work you will be expected to submit for a grade.

Research Assignments	Specific Instructions for Completing Assignment	Completion Date
Phase 1		
Phase 2		
Phase 3		
Phase 4		
Other Assignments		

What's the Problem?

Going back in time, that's the problem. Think about it for a minute. How would you study a habitat that existed over 70 million years before present? For that matter, how would you study what happened in a habitat a day ago or a few hours ago? You can not directly observe a process that has occurred in the past. However, can we use data existing in the present to make conclusions about what existed in the past?

There is another kind of exercise that is often very valuable as a process for delineating complex processes. This exercise is a thought experiment, also called a Gedanken experiment. Gedanken comes from German, and is the plural of gedanke which means thought. Thought experiments evolved in physics and were basically demonstrations or calculations that were based on components of a theory that could not be tested experimentally. In the spirit of thought experiments, we took a theoretical construction of earth 70 million years ago and made calculations of what wetlands sediments would be like if they were to transform into certain types of coal.

To make a long story short, our thought experiment led us to a design of what we would find if we were able to walk about in a Late Cretaceous coastal wetlands. Specifically, we were interested in creating all of the conditions that would allow us to study the formation of coal from plant remains that became sediments in a wetlands area. In order to do this, we needed to study what scientists have concluded about the structure of ecosystems during that period, such as the likely fauna and flora, biogeochemical processes that were occurring in certain types of habitats, and the conditions under which organic debris would enter the pathway leading to coal formation.

Now, in living color, we offer you a glimpse of our thinking in the virtual world of a Late Cretaceous mire. Recall that in the research on Yucca Mountain, we asked you to study the area in detail and begin to look at specific processes related to groundwater flow and seismic activity that might influence the stability of the site as a repository for nuclear wastes. We are sending you to the Cretaceous Mire so we can find out if it was an area in which organic matter was sedimenting and becoming a potential coal bed. In order to determine this possibility, we ask you to explore the site, collect core samples from various areas of the mire, analyze the core samples to study the depths of different kinds of materials, and, finally, construct cross-sections of the mire in order to develop a profile of the mire sediments. This sequence of research falls nicely into four phases of work.

Background Information

Peat deposits that may become coal in the geologic future are known to form from the remains of plants that grow and die in wetlands, or mires. However, a typical mire is more than just dense vegetation. These habitats are complex natural systems in a dynamic transitional environment between land and water. There are places with luxuriant plant growth, places with little growth, and places that are submerged. A mire is continually subject to flooding by fresh water from inland streams, or by salty or brackish tidewater from a nearby ocean. Floods carry mud, sand, and dissolved minerals that can mix with the plant debris and alter its composition. There may also be deposition from above, such as ash fall from volcanoes erupting upwind, or dust storms. How do all of these factors influence the fate of a peat deposit? How easy is it for coal to form at all, let alone a high-carbon, low-ash, low-sulfur coal that would be most valued for economic and environmental reasons?

The mire which you will explore is representative of the vast coastal wetlands along the shore of a shallow sea that rose over the middle of North America in the late Cretaceous Period, about 70 million years ago. This Late Cretaceous Mire is as realistic as our geological knowledge of that time allows. We know the valuable Western Interior coal resources of the Colorado Plateau originated in such environments at this particular time in the geologic past, but we also know that high-quality coal was not formed just anywhere in the region. Pay close attention to the surroundings as you travel. Note when you cross streams or other areas of open water, and whether plant growth is heavy or more sparse. In order to fully characterize this mire and determine whether it could someday yield coal deposits, you will need to drill into it in as many locations as you can.

The same simulation technology that has brought you back to the Late Cretaceous has also provided you with the latest in drilling-sampling devices for the time traveler. We call this rig the Pocket Mire-Master, and it is so sophisticated you only need to "tell" the computer where to drill with a double-click. This Pocket Mire Master has a non-invasive drill assembly to minimize any possible destruction to the future that might be done when you are "mucking about" in the Late Cretaceous.

Remember, much of this mire is unstable and hazardous. The simulation designers have identified safe locations for drilling. These areas can be identified on the view screen as you traverse the mire by a special glowing alert sign that reads "MIRE ACCESS AREA." Each stop on your path has one such access area. Make sure you use the Pocket Mire-Master at each of these locations.

Research Questions

Earlier, we said we were sending you back in time to answer a question about the potential for a particular Late Cretaceous Mire to be a starting point along the pathway to coal formation. More specifically, we could frame the question in the following manner:

> Is this late Cretaceous Mire an area in which organic matter was being deposited and becoming a potential coal bed?

We also suggested four phases of research, as shown in **Research Note 5.1**. Use the phases we proposed, or talk with your classmates and instructor to see if there are other approaches you could use to determine the subsurface structure of the Mire and its potential as a starting point for coal production.

Research Note 5.1: Phases of research for studying subsurface structure in a Late Cretaceous Mire.

1. Explore the site and examine the landforms, stream structure, and patterns of vegetation.
2. Collect core samples from various areas of the mire.
3. Analyze the core samples to study the depths of different kinds of materials in the mire and, also, the amounts and kinds of organic material that could form coal.
4. Construct cross-sections of the mire in order to develop a profile of the mire sediments and their composition.

Phases of Research

Phase 1 — Survey the Mire.

Remember the overarching question we need to answer:

> Is this late Cretaceous Mire an area in which organic matter was being deposited and becoming a potential coal bed?

All of the phases of research provide some of the data we need to answer this question. In **Phase 1** you will need to explore Late Cretaceous Mire before you begin to drill and record geologic and chemical data. To begin, click on the image of the Mire on the initial screen to enter the virtual environment. Then, before going further, you should fill in the Mire map (**Figure 5.1**) and use it to navigate through the wetland. Note that the map is organized as a grid; this scheme will be useful to you as you begin to analyze your data. Each square on the grid measures 100 meters by 100 meters. An identical map with important features is shown on the computer screen and shows your location in the Mire at all times.

Again, we note this mire is full of natural hazards and inaccessible or unstable places. Given the pitfalls of time travel, we have constrained the areas you can visit in the Mire. A safe route is marked as the dark line on the map and the red path superimposed on your computer view of the mire. This path allows you to reach and study all of the most important types of Mire features.

Task 1:
To begin, make a general survey of the Late Cretaceous Mire. Note that the Mire is located on a nearly-level coastal plain about 10 kilometers west of the seacoast. Locate and sketch the major streams that cross the mire on **Figure 5.1**. In which direction are the streams flowing?

_____
_____
_____
_____

Task 2:
Now examine the streams in more detail. Notice the banks of the streams are higher than the water level in the streams and higher than their surroundings. Explain why.

_____
_____
_____
_____

Task 3:
Locate areas of standing surface water. Sketch these areas on **Figure 5.1**. How does the water level in these places compare to the water level in the streams?

_____
_____
_____
_____

148   Chapter 5: Mire to Fire

Task 4:
Study the mire structure, and decide where you would expect to find active deposition of inorganic sediments, such as mud and sand, taking place. Why?

_____
_____
_____
_____

Task 5:
Think about water movements through the Mire. Would you expect to find significant erosion of sediments anywhere in this Mire? Why or why not?

_____
_____
_____
_____

Phase 2 — Collect and log core samples.

Now you can get down to some real data collection as you systematically move along the red path and drill at each stop. Remember to pay attention to your surroundings. The streams, areas of open water, and vegetation patterns are important elements of the Mire structure and give clues about where organic material can form sediments. You will still need to drill and take core samples that provide data on subsurface structures and deposits in order to determine whether the Mire sediments could be transformed to yield coal deposits.

Your sampling tool is the Pocket Mire-Master. Note that it is a sophisticated, non-destructive sampling device, and you only need to double-click on the site you want to sample. Safe locations for drilling can be identified on the computer screen as you traverse the mire. A special glowing alert sign that says "MIRE ACCESS AREA" will let you know you have reached a sampling site. Each stop on your path has one such access area. Make sure you use the drill at each of these locations.

Task 1:
Each time drilling is done, note the grid square (e.g., A1) where the core was obtained. Record these locations in **Table 5.1**. At the same time, mark each square drilled on the Mire map (**Figure 5.1**).

Task 2:
The other data you will need to collect are the types of organic and inorganic sediments encountered as you drill into the mire and the depths at which each layer of sediments begins and ends. These data are referred to as well logs. Record your well-log data in **Table 5.1**. Notice that the Pocket Mire Master is also showing you some other data concerning peat and ash chemistry. We will selectively return for these data in the next lab - after we understand the subsurface better!

Phase 3 — Analyze core samples.

Let's go back to the research question we need to answer:

> Is this late Cretaceous Mire an area in which organic matter was being deposited and becoming a potential coal bed?

We need the data from each phase of research in order to answer this question. In **Phase 3** we analyze the core samples and determine what materials are in the sediments, how deep they are, and how they might influence coal formation.

Task 1:
How do clay, silt, and sand differ in terms of grain size? How is this related to the nature of the processes that form sedimentary deposits?

_____
_____
_____
_____

Task 2:
What is meant by "brackish" water and peat?

_____
_____
_____
_____

Phase 4 — Construct cross-sections of the Late Cretaceous Mire.

Your drilling of core samples and analysis of the layering in these cores has hopefully given you some knowledge of the sediments in the Mire, both organic and inorganic. However, you now have cores from various sites in the Mire, which are basically snapshots of the layering in different spots. You can use these data to construct a cross-section of the Mire. A cross-section is the view you would have if you could slice across the Mire, remove one whole chunk, and look directly at the layering of sediments. To obtain a better picture of the subsurface structure, you will need to construct east-west and north-south cross-sections.

Task 1:
From the marked Mire map (**Figure 5.1**), identify where you have made the longest end-to-end traverses of the mire. One should be east-west along a single row, and one north-south along a single column. Hopefully, you have obtained at least one core in every grid square along the traverse!

Task 2:
Use the graph paper provided as **Figures 5.2** and **5.3** to construct east-west and north-south cross-sections, respectively. Remember that the scale of the Mire map is 100 x 100 meters per square. Measure the distances between each core locality and mark them off, to scale, along the top of the graph paper.

Task 3:
Represent each drill hole as a vertical line. The maximum drilling depth is 40 meters. In order to make workable cross-sections, you will have to exaggerate the vertical scale. You may choose your own horizontal and vertical scales, but make sure they are the same for both cross-sections.

Task 4:
Using the well-log data in **Table 5.1**, mark contacts between layers in each drill hole at the correct scale depths, and connect corresponding contacts between the drill holes to complete your cross-sections. You may need to make some subjective interpretations where layers "pinch out" between drill holes.

Task 5:
You have completed quite a bit work for **Tasks 1-4**. Now try to synthesize your understanding of what is happening in the Mire by answering the following questions:

a) Where are the deposits of the coarsest sediments (e.g., sand is coarser than clay) in the Late Cretaceous Mire located? Why?

b) How does the grain size of inorganic sedimentary deposits vary among different places in the mire? Explain how you would account for this.

c) Imagine you only had access to the east-west cross-section, and you were trying to estimate the continuity of the peat layer, how would the north-south cross-section temper your estimate?

In the Know about Peat and Coal

We often find that people have incomplete knowledge or misconceptions about how coal is formed and the environments which create the organic debris that can be transformed into coal. Suppose you are asked by a local school system to provide a workshop for middle school teachers. When you inquire about the science background of these teachers, you find most had several science courses when they were an undergraduate, but only a few majored in a science discipline. Very few of the teachers had environmental sciences or geology courses. Based on your experiences in the late Cretaceous Mire, use the following questions as a framework for developing the workshop.

1. Before the workshop begins, how would you assess the teachers' knowledge about coal formation and geology? How could you use this knowledge to shape your presentation?

2. What are the geological eras and their periods? During which periods were there environments that created the organic debris that would become coal?

3. What types of habitats would you find in the Late Cretaceous, including fauna and flora?

4. How would you describe the formation of different types of coal from different types of sediments in the Late Cretaceous?

5. Suppose you used the research simulation as part of the teacher workshop training. How would you describe the limitations in the thought experiment that was used to create the simulations?

6. After the workshop was completed, how would you assess the teachers' knowledge of the material you presented?

152 Chapter 5: Mire to Fire

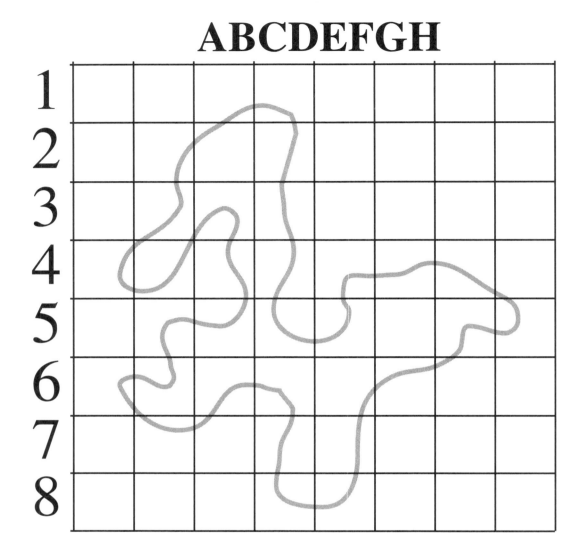

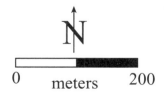

Figure 5.1

COAL POWER PROBLEM 1
MIRE LOCATION GRID

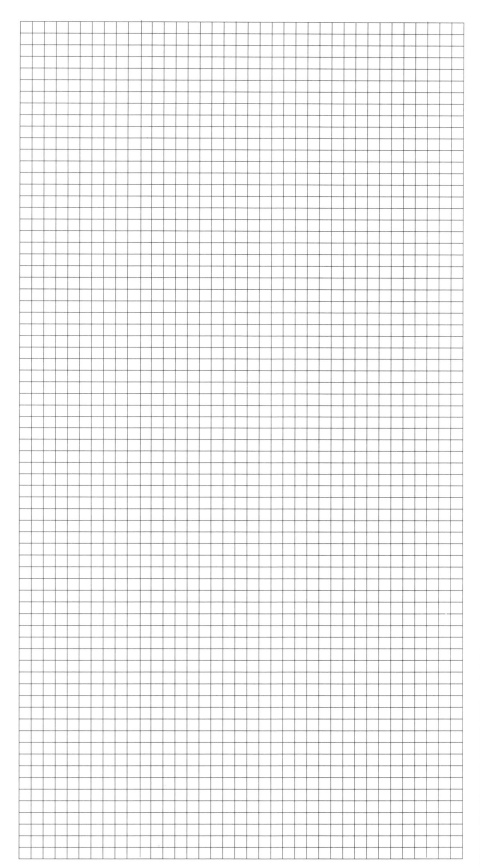

Figure 5.2

COAL POWER PROBLEM 1
NORTH-SOUTH
MIRE CROSS SECTION

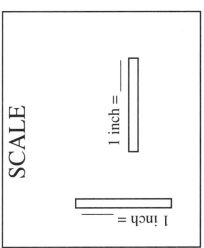

154   Chapter 5: Mire to Fire

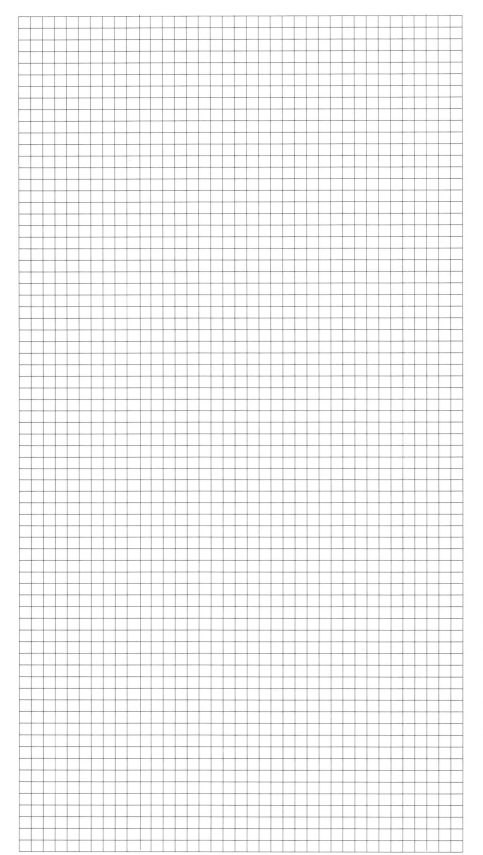

Figure 5.3

COAL POWER PROBLEM 1
EAST-WEST
MIRE CROSS SECTION

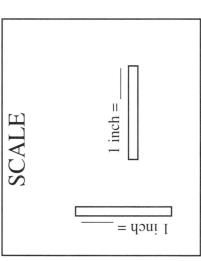

Coal Power Research Problem 2: Predicting Coal Quality

Introduction and Advice

Before you start the research, examine **Study Notes 5.4** and **5.5** for some guidance about what you need to know before you start and what you should know when you are done. In **Study Note 5.6**, we provide a table you can use to keep track of assignments and their due dates.

Study Note 5.4: As you begin the research, you will need to have some prior knowledge in the following areas:

1. A basic understanding of Late Cretaceous environments.
2. Knowledge about the formation of sediments that can be transformed into coal through geological processes.
3. Some knowledge about coal formation, especially the different types of coal and the specific processes that lead to these types.
4. How to relate structure of ancient peat beds to quality of coal formed from these beds.

Study Note 5.5: What you should be able to do after completing the research.

1. Interpret variation in peat composition in terms of environmental variations, specifically detrital origin of ash and marine influence on sulfur content.
2. Estimate ash and sulfur contents of coals that could form from different types of peat.
3. Relate data on peat composition to the quality of coal formed from the peat beds.
4. Interpret compositional and grain-size variations among sedimentary facies in terms of transport.
5. Understand the time-transgressive nature of a mire environment (and other sedimentary environments, by extension).

> **Study Note 5.6:** Use the table provided below to list the research assignments given by your instructor. Make notes on this table about the types of work you will be expected to submit for a grade.

Research Assignments	Specific Instructions for Completing Assignment	Completion Date
Phase 1		
Phase 2		
Other Assignments		

What's the Problem?

In the previous research simulation you constructed a three-dimensional view of the Late Cretaceous Mire. From your core samples and analyses of core layers, you discovered some of the sedimentary processes controlling the formation of peat. Now, you will explore whether any of the peat you discovered and mapped previously is a chemically potential candidate for coal. However, even if the peat is a potential coal-forming deposit, there are a many reasons why we cannot be certain of the fate of this Late Cretaceous Mire. After all, there are 70 million years or so between the peat formation and our present! A lot can happen in that amount of time. Here are a few processes that would disrupt the pathways from peat to coal.

- A warmer climate, or tectonic subsidence of the region, could bring the nearby sea inland to erode away the peat deposits. A similar thing could happen with increased streamflow from the landward side.

- Should the climate become drier, plant growth could slow or even cease, and the peat layers already deposited could be exposed to the air, dry out, and decay.

- Cataclysmic volcanic eruptions, which have occurred repeatedly in geologically-active western North America since the Cretaceous, could suffocate the mire in layers of silicate ash.

However, let us assume that deposition of plant matter will continue for a time, and then the peat layers will be buried under inorganic sediment as the depositional environment gradually changes. This Late Cretaceous Mire could be on its way to becoming a minable coal deposit, but will the coal be worth mining?

Background Information

In order to complete this research project, you will first need to know some basic information about coal:

• The meaning of each of the values obtained in a proximate analysis.
• The maximum allowable ash content, in percent by mass.
• Typical sulfur contents of "low-sulfur" and "high-sulfur" coals, in percent by mass.

Think about the process you need to implement. First of all, you have quite a bit of data in **Table 5.1**. Next, you will have to find areas of peat in the Mire. Using **Table 5.1** and **Figure 5.1** as a guide, you can enter the Late Cretaceous Mire, and identify those areas of the grid that mark places where there is a peat layer. Now, you will have to sample the peat. This can be done by redrilling cores for each location that has peat. After each sample is taken, record the proximate analysis data displayed by the Pocket Mire Master in **Table 5.2**. The proximate analysis is the list of the critical constituents of peat. You may also have some samples in which you find a volcanic ash layer. When this occurs, you will need to record these data using **Table 5.3**.

The peat constituents and the amounts of inorganic materials (like ash) will influence the quality of coal that can be formed from the peat sediments. The proximate analyses you are recording provide data on the in situ peat, that is the peat on site in the ground. As you might imagine, the peat is really wet — it contains a lot of moisture. To compare the energy potential of each peat you have to compare the dry masses of peat to each other. The final step in your process should be to recalculate the composition of all of your peat samples to get a "moisture-free" estimate. This is done by summing the non-moisture parameters of **Table 5.2** to get a total, then dividing each parameter by the total to get the "dry" or moisture-free percent of each parameter by weight. Record your moisture-free percentages in the "Total dry" column of **Table 5.2**.

Think through the process we have outlined above. It is crucial that you have an understanding of the steps in the process and the rationale for completing these steps. Remember, we are interested in the quality of the coal that can be formed from the peat. Of course, different areas of the Mire have peat with different compositions. Since these compositions influence the quality of the peat as potential coal, we need to study differences in composition and try to determine what quality of coal could be formed from the different areas of the Mire.

Research Questions

The research in this simulation is very descriptive in nature. We recommend the following overarching research question as a guide for conducting the study.

> What is the relationship between peat composition and potential coal quality?

You can see this is a broad question. However, in terms of the research we can conduct in our Late Cretaceous Mire, there are two subordinate questions that together will help us answer our major question. First, we ask, "What is the composition of the peat in different regions of the Mire?" This question can be answered by the proximate analyses you will be conducting. Second, we ask, "How can we interpret the Mire as a potential coal forming environment?" This question requires you to study data related to the Mire's structure and the processes that are likely to shape the sediments in the Mire.

The major research question and its two subordinate questions provide a framework for designing and implementing a study of the Mire. As in other simulations, we have to answer the subordinate questions in order to answer the overarching question. We feel the subordinate questions each fit into a phase of research. Thus, we propose two phases in the study. First, we want to study peat composition in different regions of the Mire. This phase can be simply called "Predicting coal quality." Second, we are interested in the processes shaping the Mire and its sediments. Again, for simplicity, we can call this phase "Interpreting the Mire environment." As for our other research simulations, we encourage you to work with your faculty member or classmates to refine the question, and so plan and refine the phases of research you conduct.

Phases of Research

Phase 1 — Predict coal quality.

We want to focus on the composition of peat and try to understand how composition relates to potential coal quality. Using **Table 5.1** and **Figure 5.1** to guide you, enter the Late Cretaceous Mire Simulation and redrill the grid areas that have a peat layer. For each location that has peat, record the proximate analysis data displayed by the Pocket Mire Master in **Table 5.2**. The proximate analysis is the list of the critical constituents of peat. If the peat has a volcanic ash layer record its composition in **Table 5.3**.

The proximate analysis you are recording is for the in situ peat - the peat in the ground. As you recall, the peat is really wet — it contains a lot of moisture. Again, remember that we want to compare the energy potential of each peat. You will have to recalculate the composition of each of your peat samples on a "moisture-free" basis and record your calculations in the proper places in **Table 5.2**. As we mentioned in the Background section, this is done by summing the non-moisture parameters of **Table 5.2** to get a total, then dividing each parameter by the total to get the "dry" or moisture-free percent of each parameter by weight. The sum of the moisture-free percentages provides you with an estimate of "Total dry." Record the sum in **Table 5.2**.

Task 1:
Consider the factors that shape moisture content of peat deposits. What will probably happen to the moisture in these peat deposits if they become deeply buried and subjected to rock-forming pressures and temperatures for millions of years?

Task 2:
Think about the ash and sulfur content of the peats. How could these change in the same circumstances as above? Explain your answer.

Task 3:
Identify which of the peat layers you sampled have the potential to become low-ash, low-sulfur coal, and which do not.

Chapter 5: Mire to Fire

Task 4:
Actual upper Cretaceous coals mined today on the Colorado Plateau typically contain near-equal percentages of volatile matter and fixed carbon. Take a look at your recalculated analyses. Do they reflect this? If not, suggest an explanation.

_____
_____
_____
_____

Phase 2 — Interpret the Mire environment.

Task 1:
What is the origin of ash in peat? Where in the Late Cretaceous Mire are the lowest-ash peats being deposited, and why?

_____
_____
_____
_____

Task 2:
Think about the sulfur in the peat. What is the origin of high sulfur content in peat? Where in the Late Cretaceous Mire are the lowest-sulfur peats being deposited, and why?

_____
_____
_____
_____

Task 3:
Examine the composition of the volcanic-ash layer. Does this composition vary across the mire? Explain.

_____
_____
_____

Task 4:
Look again at the ash layer. Do you find the volcanic-ash layer appears to have any effect on the chemistry of the peat above or below it? Why or why not?

_____
_____
_____

Task 5:
Now, think about the transformation of peat to coal. What kind of effect might the ash layer have on coal beds if they form in the geologic future?

Task 6:
Describe what might happen to peat deposition and peat quality in this mire in the following scenarios:

a) The sea-level rises, the streams become tidal channels, and the sea gradually encroaches over the mire from the east.

b) The sea-level falls and the sea retreats farther from the mire than it is now.

c) The two streams flood, spilling over their banks. The area between them is completely inundated, and it becomes covered with standing water.

d) The water table drops below 50 meters.

e) A minor volcanic eruption somewhere to the west drops about 5 centimeters of volcanic ash over the entire mire.

Chapter 5: Mire to Fire

Task 7:
Finally, a present-day coal geologist is mapping a coal seam that is exposed along several kilometers of a hillside. Rock layers immediately above and below the seam are also exposed. Might there be any geologic clues that would indicate whether this coal is of good quality?

_____
_____
_____
_____

In the Know about Coal Quality

We recommend you conduct a literature research project. The goal of this project is to relate your findings on coal quality to the pros and cons of using coal as a fuel for generating heat and electricity. Take the point of view of a public interest research group that wants to complete an analysis of coal use in order to produce an unbiased educational packet. The packet will be distributed in a region that has a coal-fired power plant owned by a utility, which is considering building a second coal-fired plant and increasing rates for electrical service.

You are assigned a portion of the literature review which focuses on the questions below. Use these questions to guide your search of the literature. Write short answers to the questions that will be used to develop substantive, but accessible, educational materials for public distribution.

1. How is coal formed, and why does it contain materials that may have negative impacts on the environment?

2. Where are the major coal beds located, and how are these used by regional populations?

3. In the Peoples Republic of China, where a large percentage of electricity is generated by coal-fired power plants, what are the documented negative and positive impacts on economic stability, human health, and the regional environments around the power plants and mining areas?

4. What are some of the consistent patterns in degradation of human health and environmental systems from mining, transporting, and burning coal?

5. What are the benefits from burning coal as a source of fuel for generation of electricity?

6. Why are the efficiencies of coal-burning power plants different in different countries?

7. How can coal-fired electricity generation be made more efficient while not greatly increasing costs of electricity?

TABLE 5.1 Sediment Logs

Core Locality	row ___ column ___		row ___ column ___	
ENVIRONMENT	Sediment	top (depth m.)	Sediment	top (depth m.)
First unit				
Second unit				
Third unit				
Fourth unit				
Fifth unit				
Sixth unit				
Seventh unit				
Eighth unit				

Core Locality	row ___ column ___		row ___ column ___	
ENVIRONMENT	Sediment	top (depth m.)	Sediment	top (depth m.)
First unit				
Second unit				
Third unit				
Fourth unit				
Fifth unit				
Sixth unit				
Seventh unit				
Eighth unit				

TABLE 5.1 Sediment Logs

Core Locality	row ____ column ____		
ENVIRONMENT			
	Sediment	top (depth m.)	
First unit			
Second unit			
Third unit			
Fourth unit			
Fifth unit			
Sixth unit			
Seventh unit			
Eighth unit			

Core Locality	row ____ column ____		
ENVIRONMENT			
	Sediment	top (depth m.)	
First unit			
Second unit			
Third unit			
Fourth unit			
Fifth unit			
Sixth unit			
Seventh unit			
Eighth unit			

Core Locality	row ____ column ____		
ENVIRONMENT			
	Sediment	top (depth m.)	
First unit			
Second unit			
Third unit			
Fourth unit			
Fifth unit			
Sixth unit			
Seventh unit			
Eighth unit			

Core Locality	row ____ column ____		
ENVIRONMENT			
	Sediment	top (depth m.)	
First unit			
Second unit			
Third unit			
Fourth unit			
Fifth unit			
Sixth unit			
Seventh unit			
Eighth unit			

TABLE 5.1 Sediment Logs

Core Locality	row	column			column		
ENVIRONMENT		Sediment	top (depth m.)		Sediment	top (depth m.)	
	First unit						
	Second unit						
	Third unit						
	Fourth unit						
	Fifth unit						
	Sixth unit						
	Seventh unit						
	Eighth unit						

Core Locality	row	column			column		
ENVIRONMENT		Sediment	top (depth m.)		Sediment	top (depth m.)	
	First unit						
	Second unit						
	Third unit						
	Fourth unit						
	Fifth unit						
	Sixth unit						
	Seventh unit						
	Eighth unit						

Core Locality	row	column			column		
ENVIRONMENT		Sediment	top (depth m.)		Sediment	top (depth m.)	
	First unit						
	Second unit						
	Third unit						
	Fourth unit						
	Fifth unit						
	Sixth unit						
	Seventh unit						
	Eighth unit						

Core Locality	row	column			column		
ENVIRONMENT		Sediment	top (depth m.)		Sediment	top (depth m.)	
	First unit						
	Second unit						
	Third unit						
	Fourth unit						
	Fifth unit						
	Sixth unit						
	Seventh unit						
	Eighth unit						

TABLE 5.1 Sediment Logs

Core Locality	row column		row column	
ENVIRONMENT	Sediment	top (depth m.)	Sediment	top (depth m.)
First unit				
Second unit				
Third unit				
Fourth unit				
Fifth unit				
Sixth unit				
Seventh unit				
Eighth unit				

Core Locality	row column		row column	
ENVIRONMENT	Sediment	top (depth m.)	Sediment	top (depth m.)
First unit				
Second unit				
Third unit				
Fourth unit				
Fifth unit				
Sixth unit				
Seventh unit				
Eighth unit				

TABLE 5.1 Sediment Logs

Core Locality	row ____	column ____		
ENVIRONMENT				
		Sediment	top (depth m.)	
First unit				
Second unit				
Third unit				
Fourth unit				
Fifth unit				
Sixth unit				
Seventh unit				
Eighth unit				

	row ____	column ____		
		Sediment	top (depth m.)	
First unit				
Second unit				
Third unit				
Fourth unit				
Fifth unit				
Sixth unit				
Seventh unit				
Eighth unit				

Core Locality	row ____	column ____		
ENVIRONMENT				
		Sediment	top (depth m.)	
First unit				
Second unit				
Third unit				
Fourth unit				
Fifth unit				
Sixth unit				
Seventh unit				
Eighth unit				

	row ____	column ____		
		Sediment	top (depth m.)	
First unit				
Second unit				
Third unit				
Fourth unit				
Fifth unit				
Sixth unit				
Seventh unit				
Eighth unit				

Table 5.2
Peat Composition

Locality			Moisture	Volatiles	Fixed C	Sulfur	Ash	Total dry
row_____		wet						
col _____		dry						
row_____		wet						
col _____		dry						
row_____		wet						
col _____		dry						
row_____		wet						
col _____		dry						
row_____		wet						
col _____		dry						
row_____		wet						
col _____		dry						
row_____		wet						
col _____		dry						
row_____		wet						
col _____		dry						
row_____		wet						
col _____		dry						
row_____		wet						
col _____		dry						

Table 5.2
Peat Composition

Locality			Moisture	Volatiles	Fixed C	Sulfur	Ash	Total dry
row_____		wet						
col _____		dry						
row_____		wet						
col _____		dry						
row_____		wet						
col _____		dry						
row_____		wet						
col _____		dry						
row_____		wet						
col _____		dry						
row_____		wet						
col _____		dry						
row_____		wet						
col _____		dry						
row_____		wet						
col _____		dry						
row_____		wet						
col _____		dry						
row_____		wet						
col _____		dry						

Table 5.2
Peat Composition

Locality			Moisture	Volatiles	Fixed C	Sulfur	Ash	Total dry
row_____		wet						
col _____		dry						
row_____		wet						
col _____		dry						
row_____		wet						
col _____		dry						
row_____		wet						
col _____		dry						
row_____		wet						
col _____		dry						
row_____		wet						
col _____		dry						
row_____		wet						
col _____		dry						
row_____		wet						
col _____		dry						
row_____		wet						
col _____		dry						
row_____		wet						
col _____		dry						

Coal Power Research Problem 3: Locating a Coal-Fired Power Plant

Introduction and Advice

Before you start the research, examine **Study Notes 5.7** and **5.8** for some guidance about what you need to know before you start and what you should know when you are done. In **Study Note 5.9**, we provide a table you can use to keep track of assignments and their due dates.

Study Note 5.7: As you begin the research, you will need to have some prior knowledge in the following areas:

1. Ability to interpret and construct topographic maps, simple geologic maps, and simple geologic cross-sections.
2. Basic algebra skills.
3. Experience synthesizing related data sets.

Study Note 5.8: What you should be able to do after completing the research.

1. Interpret geologic data from an area and be able to organize these data into a three-dimensional representation of the area.
2. Create a geologic cross-section of an area.
3. Combine related data sets into a problem-solving framework for an environmental/economic problem.
4. Understand the multiple factors shaping optimization of resource locations and cost effectiveness.
5. Understand the complex relations and constraints in siting an industrial facility.

Chapter 5: Mire to Fire

> **Study Note 5.9:** Use the table provided below to list the research assignments given by your instructor. Make notes on this table about the types of work you will be expected to submit for a grade.

Research Assignments	Specific Instructions for Completing Assignment	Completion Date
Phase 1		
Phase 2		
Phase 3		
Phase 4		
Phase 5		
Other Assignments		

What's the Problem?

You know that most Americans use electricity in a large percentage of their daily activities. The enormous demand for electricity requires that utility companies constantly re-evaluate their available power. They must plan for tapping into power grids supplied by distant power plants, or plan to modify their existing power plants and build new plants in order to meet the needs of the populations they serve. Again, we challenge you to find out where your electricity is generated. Also, think about how much you use, and how you might estimate usage by the city or town where you live.

Remember that in the United States 50-60% of the total electricity is generated by power plants that use coal. Where are these plants located? How were the sites chosen? Should the sites be close to the populations for which they provide electricity? Or are there other factors that shape decisions about where to locate coal-fired power plants? Once again, is it likely that at least some of the power you use comes from a coal-fired power plant? How would you find out?

In the previous two simulations, you had the opportunity to study coal formation, starting with deposits in a Late Cretaceous Mire that had the potential for becoming coal. Let's get out of the past and come to the present time, some 70 million years after the Mire. Let's suppose that some of the Mire deposits have been transformed into coal. We can now think about the quality of the coal and if it can be used to power a coal-fired plant that will generate electricity for a regional population. As you will see, the research we conduct in this simulation is a little different from our earlier studies. Our work in this virtual world will be focused on problem-solving related to locating a power plant that can utilize a coal bed.

For the moment, put aside issues related to the pros and cons of using coal for generating electricity, and put aside the host of environmental issues related to mining, transporting, cleaning, and using coal. We are asking you to accept that we have a specific assignment, namely to study an area and locate a coal-fired plant in that area. Use this simulation to learn about the process of siting power plants, and use that knowledge to become better critical thinkers about all of the issues involved in using coal for generating electricity.

Background Information

You have a unique job that is a combination of civil engineer, geologist, lawyer, planner, and contractor. A utility company hires you to locate an area that will support the needs of a 720 megawatt Coal-Fired Power Plant. To accomplish this goal, you will have to explore the availability of the resources in the area necessary to run a coal power plant and then make a recommendation for the power plant site where these resources can be optimally utilized.

As is usual in the real world, there are constraints on your firm:

- Money — You have a limited budget for studying the areas resources and synthesizing the data in order to make a recommendation.
- Type of power plant — The plant is a 720 megawatt power plant that must operate for 20 years. You must locate the plant near enough to the coal mine so transportation costs are minimized. This kind of setup is known as a mine-mouth operation. That is, the mine is close to the mouth of the ever hungry power plant. You will also need a coal supply system and water for cooling

- Type of coal mine — The coal mine cannot exceed 70 meters in depth. This precludes any underground mining. All the coal must be extracted from an open pit. Also, the mine must have enough coal of an acceptable quality that the power plant can use the coal as fuel for 20 years.

Just to give you a sense of the scope of the problem, let's think about the amount of coal it would take to run a power plant for 20 years. We said that we want to site a 780 megawatt power plant. In order to operate such a plant for 20 years, you will need about 63 million tons of subbituminous coal (density of 1400 kg/m^3). Now, we can calculate the total volume of coal needed:

(63,000,000 tons $\times$ 4,410 kg/ton) / 1400 kg/m^3 = **198,450,000 m^3 !**

The above volume gives you a sense of the amount of earth that would have to be removed from the mined area. In other words, a whole lot! To give you some perspective, this volume is equal to a cube of coal more than 580 meters on a side, or a slab of coal about 30 meters thick, 13 kilometers long and half a kilometer wide!

There are other complications. For example, how much water will we need for cooling? The plant we propose will use 500,000 gallons per minute. Either we will need a large river or we will have to locate a water storage and cycling facility near the plant. For this simulation, we have chosen an area which does not have a river with sufficient flow such that its natural discharge can be used to meet the plant's needs without unacceptable environmental impact on the river itself. You will have to look at available water sources, and also at landforms that can be used as natural reservoirs.

You get the picture. There are a lot of factors to consider. We constructed the simulation software to display a shaded relief base map of the area in which we want to site the plant. Overlaying the map are a series of grids corresponding to the various forms of information you will be collecting. Clicking on a cell will allow you to study the area of that cell, including subsurface geology, topography, water availability, and land use. Your budget allows you to conduct research on a limited portion of the region. Thus, you have to make your selections carefully. You will work through the different databases in order of importance to this project: (1) identifying the coal supply, (2) looking for a reservoir location, (3) examining surface water rights, and (4) analyzing patterns of land use (including human population distributions).

Research Questions

We said above that this simulation provides a different kind of research opportunity. You might argue that we do have a simple research question, namely:

Where do we site a 720 megawatt coal-fired power plant?

The subordinate questions could focus on identifying the locations of coal seams, landforms for cooling reservoirs, choosing water systems to fill the reservoir, and finding areas where changes in land use will not be controversial. The pattern of asking a major question and breaking that question into subordinate questions is the approach we have taken in other simulations.

However, in this simulation, we want you to take a very applied approach. You should still drive your research by focusing on data you need in each of the resource areas essential to running the plant.

But instead of asking specific research questions, try to create a strategy for approaching the problem. You will find that in this simulation you have much more freedom for setting up sampling methods. Use this freedom to think about what you want to know and how you can conduct your study in order to get the information you need. Also note that in a manner similar to what happens in the real world, your studies will cost money and be limited by your budget.

Phases of Research

Phase 1 — Finding the coal supply.

In this phase your primary goal with the geologic database is to locate the coal-bearing unit in the area and understand its geometry. Thus, you will be able to calculate the volume of the coal available and determine if it is sufficient to support 20 years of power plant operation.

Task 1:
Click on the landscape next to the "Mine" signpost on the inital screen and then select the **Geology Database** from the buttons below on the map. This will allow you to estimate the volume and tonnage of available coal. However, don't just plunge in. Devise a strategy for your exploration. Begin by focusing your attention with a few questions about the rocks and topography. For example, click the geology button and take a look at the stratigraphic section that is displayed on the left side of the computer screen. The sedimentary rocks of Cretaceous age dominate this area. Imagine these rocks, which have varying degrees of hardness, in different orientations and try to answer the following questions.

a) What kinds of landforms would you expect if these rocks were vertical (on edge)?
___
___
___

b) What kinds of landforms might you see if they were horizontal (flat lying)?
___
___
___

Task 2:
Look at the topography shown in the simulation, and take in the big picture. Identify any mountains, ridges, valleys, or arroyos. Identify any rivers or streams. Describe any patterns or linear features that might tell you about the structure of the rocks beneath the surface. What kinds of landforms or features dominate the topography?
___
___
___

Chapter 5: Mire to Fire

Task 3:

A relationship does exist between the orientation of the sedimentary rocks and the form and expression of the topographic features. You can use this relationship to guide you in your exploration. What connections do you think might exist between the dominant landforms and the way the sedimentary rocks might be oriented?

_____
_____
_____

Task 4:

There are sixty possible geology cells to explore and you have a budget to look at 20 of them. First, outline your strategy for exploration. Then use the table below to indicate the first 10 cells you will choose, and in what order. Take a look at **Research Note 5.2** before your start. Remember, you can change your mind later, and revise your strategy and cell choices to help you learn more.

Research Note 5.2: Selecting geology cells.

Every time you select a cell (by clicking on it) the following events happen:

1. Your field crew will map the surface geology for the cell.
2. The drilling contractor will supply you with a core log of the sub-surface geology.
3. You've spent $ 50,000 !

---

a) Strategy:

_____
_____
_____
_____
_____
_____

b) List the cells you want to sample by indicating their row and column on the map. Also, place the cells in the order you want to sample them.

	1	2	3	4	5	6	7	8	9	10
Row										
Col.										

Task 5:

Each time you explore a new geology cell follow these steps:

a) The surface geology should be sketched onto your geologic map (**Figure 5.4**) in the appropriate cell. Choose a color or pattern for all the Quaternary beds, and map them together as one color or pattern.

b) Did you find any coal? If so, can you predict where the coal bearing bed leads from this cell? Where do you think you will find it next, and why?

c) If you didn't find any coal, does the geology tell you anything about which direction to go to find coal? For instance, if you found sedimentary rocks on the surface that are stratigraphically below the coal and the beds are dipping towards the northwest, this would tell you the coal might be found at the surface northwest of the current cell. How does your analysis of the cell data effect your strategy for cell exploration?

d) Look at your core log. Was coal detected on the log? If so, at what depth? These core logs will also be used later to produce cross-sections. How does your analysis of the core data effect your strategy for cell exploration?

Task 6:
You can now determine the volume of available coal. If you have exhausted all of your funds for geologic exploration, and/or have a clear idea of where the coal is above and below ground, then you should begin making cross-sections. The cross-sections will help you estimate the available coal, and when they are complete, you should have an idea of the volume and tonnage of available coal.

Look at the cross-sections (**Figure 5.5**). These are grouped into east-west and north-south cross-sections of the simulation area. Choose at least two of each kind, and fill in the geology (as best you can) based on the surface map you have and the core logs which can be accessed from the geology database by clicking on the Core Log button. The core logs are located in the center of each cell. As you fill in your cross-section, link each core together to show what happens to the beds between cores.

In some of your cross-sections that are perpendicular to the regional strike direction, you should see the coal beds tilted and dipping from the surface. Recall that any coal below 70 meters is beyond your concern. You should be able to use this depth to draw a line that marks the lower limit of the accessible coal. This line along with the top and bottom contacts of the coal bed and the ground surface should form a closed shape (a parallelogram) that represents the cross-sectional area of the coal bed. Estimate this cross-sectional area as best you can.

   Cross-sectional area of the coal in m2 (square meters) _____

How far does the coal bed extend along its strike? Measure or estimate this length and multiply it by the cross-sectional area to calculate the volume of the coal bed.

   Continuous coal bed volume in m3 (cubic meters) _____

Is this enough coal for 20 years?

Phase 2 — Locating a reservoir to store water for plant cooling.

We can now use the simulation software to select the **Reservoir Database**. The power plant will require an enormous amount of water for cooling. Although this water is not used up in the process, it is heated up as it passes through the plant. A large reservoir will be needed to both store the water and allow it to cool before being cycled through the plant again. The reservoir will have to be filled from the nearby river system since groundwater is unavailable. This reservoir must be next to the plant to keep pumping costs at a minimum. Let's get a "mental" picture of the reservoir, and outline the processes affecting the flow of water in and out of the reservoir.

Chapter 5: Mire to Fire

Ideally the reservoir should be an elongate wedge-shaped hole that will hold at least 10 times the daily requirement of water pumped through the plant (estimated to be about 720 million gallons). The discharge from the plant will occur at the shallow end (water temp = 1000 F), while the plant intake will be at the deeper end (water temp = 750 F). The water cools by evaporation as it flows from one end to the other. Expect some water loss from evaporation and designate this loss: W_{evap}. This loss is dependent on the surface area of the reservoir.

The costs to excavate a hole this size would be enormous. Instead, an engineer would ideally locate an existing drainage that could be dammed and flooded. This of course will drown the habitat found in the drainage but will create a new wetland. Expect some water loss from the floor and sides of the reservoir into the surrounding rocks. Designate this water loss: W_{leak}.

The plant will lose some water during the scrubbing process where sulfur dioxide gas is "washed out" out of the furnace gas. Designate this water loss: W_{srb}. For this power plant the design engineers have estimated W_{srb} to be about 1000 gpm.

If the reservoir is full, we need to have a constant flow from the river (designated Wriv) to maintain the reservoir. As it turns out, Wriv will also be sufficient to initially fill the reservoir while the plant is being constructed over several years.

$$W_{riv} = W_{evap} + W_{leak} + W_{srb}$$

Task 1:
Think about a strategy. The goal is to find potential reservoirs (existing drainages that could be dammed then flooded). These have to be close to the coal bed we located earlier. Refer to **Research Note 5.3** to get some information on selection of areas to study.

Research Note 5.3: Selecting cells on the map grid to study reservoir locations.

Every time you select a grid cell (by clicking on it):

1. Your survey crew will map the reservoir potential for the cell. If part of the drainage has the potential to be dammed up, it will be designated a special color.
2. Your crew will also tell you the values for W_{evap} and W_{leak} for that particular reservoir.
3. Depending on the reservoir, the crew will spend $10,000 per cell, and with a budget of

---

Each time you explore a new cell follow these steps:

a) Record the values for W_{evap} and W_{leak} on your reservoir map (**Figure 5.6**).

b) Calculate the value for W_{riv} and record it on the reservoir map. (**Figure 5.6**).

Phase 3 — Studying sources of surface water.

We can now study the area, and see if there are suitable sources of surface water that can be used to fill up the cooling reservoir. The local river system will be accessed to supply water that will be "stockpiled" in the reservoir. This requires you to identify sufficient surface water rights (volume of water legally available to pump out of a river or stream). For this study, select the **Surface Water Database.**

Task 1:
Again, plan a strategy for studying the area. How will you decide where to look for surface water rights? Water rights exist in surface waterways such as in streams and in rivers. Generally, there is always a protected water right inherent to any river or stream to ensure a constant base flow. This prevents the stream or river from being pumped below the level required to sustain the aquatic life of the waterway.

If you observe the topography in the simulation, you will notice there appear to be at least two rivers shown on the map. These would be good places to search for water rights. To fine tune this search recall that you know from the previous section where your potential reservoirs are located. The closer the surface water right is to both the reservoir and the coal bed the lower the costs for pipelines and pumping will be. Consult **Research Note 5.4** for pertinent information.

Research Note 5.4: Selecting cells on the map grid to study surface water locations and rights.

Every time you select a grid cell (by clicking on it):

1. Your legal crew will report the available water rights for the cell.
2. You will spend $10,000 per cell.

Task 2:
As you work through your study of the surface water, you will need to record the water rights value on your surface water map. Use **Figure 5.7** for this purpose.

Phase 4 — Acquiring land for the power plant site.

You should have some areas in mind for the placement of the power plant based on your work in previous sections. What kinds of activities occur on those pieces of land? Last, but certainly not least, are people and where they live. Will they be willing to move?

Task 1:
First, examine land use patterns. Use the simulation software to select the **Land Use Database**. There are a variety of land uses in the area you are studying. For example, consider the following:

- Commercial Services, Industry, and Light Industry — Land with these uses are normally too expensive to purchase, because established businesses would require relocation at enormous costs.

- Shrub — Brush range and mixed range provide habitat for many species of plants and animals. Cattle are run periodically through these areas for grazing. Both of these ranges are generally affordable.

Your strategy for selecting potential sites for the power plant will require these sites look promising based on your geologic, surface water, and reservoir studies. Refer to **Research Note 5.5** for some guidance.

> **Research Note 5.5: Selecting cells on the map grid to study land use patterns.**
>
> Every time you select a grid cell (by clicking on it):
>
> 1. Your planner crew will report the land use for the cell.
> 2. You will spend $5,000 per cell.

As you explore each new cell, be sure to record the use on your land use map in the appropriate cell. Use **Figure 5.8** for this purpose.

Task 2:
Select the Population Database using the simulation software. Usually, it would be prohibitively expensive to buy an established town, relocate the people, tear down the town buildings, and construct a power plant in its place. Sometimes, sparsely populated regions might have residents who are willing to sell out and move. Certainly, regions with no human inhabitants are more attractive to industry when they try to acquire land.

You will need to identify those places that look promising. Remember, you have to take into account all the important factors for siting the plant: geologic, reservoir sites, surface water availability, and land use. **Research Note 5.6** provides the usual constraints.

> **Research Note 5.6: Selecting cells on the map grid to study population distributions.**
>
> Every time you select a grid cell (by clicking on it):
>
> 1. Your survey crew will report the population for the cell.
> 2. You will spend $2,500 per cell.

Each time you explore a new grid cell make sure you record the population value on your population map. Use **Figure 5.8** for this purpose. If two or more population densities appear in a cell, you have to decide on a method that gives an average population value for the cell.

Phase 5 — Synthesis of data from coal supply, reservoir location, surface water availability, land use, and population distribution studies.

Now that you have obtained a reasonable set of data, how do you locate a site for a power plant? You are looking for a site near the necessary resources to operate a coal-fired power plant, and where these resources are 'optimized'. For the power company, this optimization translates into costs. Basically, the power company wants to know which site will require the lowest capital investment to construct and maintain the coal-fired power plant? In **Research Note 5.7**, we provide a framework you can use for problem-solving and deciding on a reasonable location for the power plant. We recommend you write short technical papers in response to each item in **Research Note 5.7**, and then use your findings to recommend a site for the plant. Put the whole set of papers together into one final report, supporting your written work with maps and tables from your studies..

Research Note 5.7: Sections of a technical report that will make a recommendation to the power company about location for the proposed power plant.

1. Coal Source — Describe the location of the coal seam and the extent of the seam in terms of both the area below which the seam is located and the total volume of usable coal. Be sure to report calculations and methods for estimation of area and volume.

2. Water Issues — Evaluate possible reservoir locations and potential water rights that match the requirements of the reservoir

3. Land Use Constraints — Examine the important constraints on possible sites that are related to current land use and population densities

4. Recommendation — Justify your choice of site, including direct connections to the data you collected during this study.

In the Know about Where the Power Plant Will Go

Geographical Information Systems (GIS) are a category of computer software which provide a way of layering data for a region. For example, a GIS allows you to create a map of an area and then create multiple data layers for that map. Consider these possible data layers:

- Topography
- Subsurface geology
- Surficial geology
- Distribution of surface water
- Distribution of groundwater
- Distribution of vegetation
- Land use patterns
- Human population density
- Precipitation

In a simple sense, you enter data into the computer in a way that it creates multiple data bases for different kinds of variables at every point on a mapped surface. Of course, in the real world, there are many, many processes going on at any point on the earth's surface, and so innumerable data exists which you could collect. A GIS allows you to store multiple kinds of data (e.g., topography, subsurface geology, average precipitation) at every point on a map. Therefore, this kind of data layering system can partially represent the multiple data existing at a site in the real world corresponding to the site on the map. We say "partially" here, because there are so many data at any point on earth and because every place on earth changes through time.

Go to the library and read about Geographical Information Systems. Then answer the following questions.

1. Why were Geographical Information Systems (GIS) developed?
2. How did the research simulation you just completed mimic a GIS system? What were the differences between the simulation and most GIS software?
3. If you were assigned the task of siting a coal-fired power plant, how could you use a GIS to store and analyze your data?
4. How would you use a GIS to represent the layered data of a 2,000,000 m2 site over the period of 10 years, with monthly sampling of 30 variables at each of 100 locations within the site?

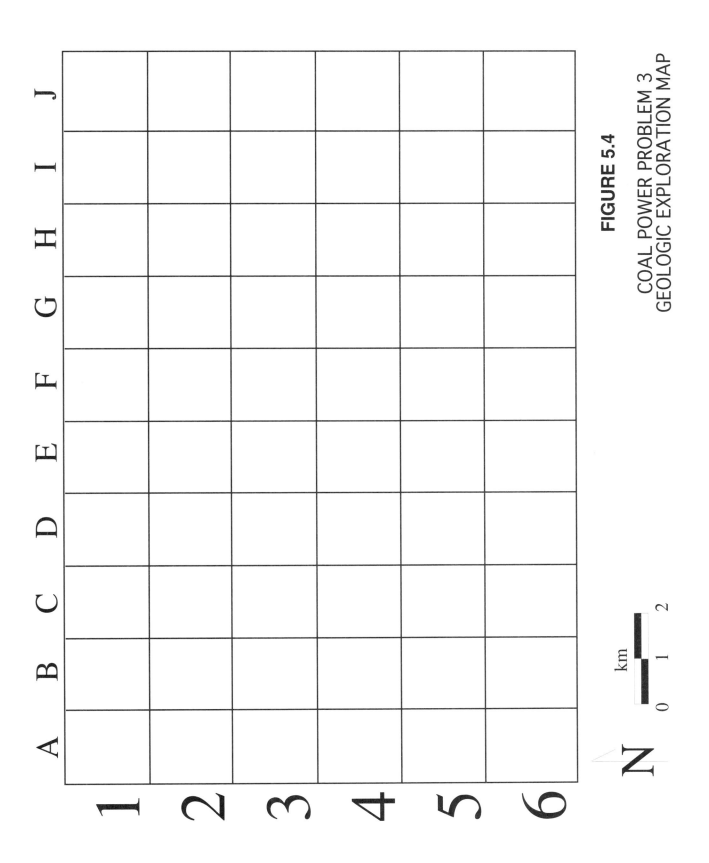

FIGURE 5.4
COAL POWER PROBLEM 3
GEOLOGIC EXPLORATION MAP

184 Chapter 5: Mire to Fire

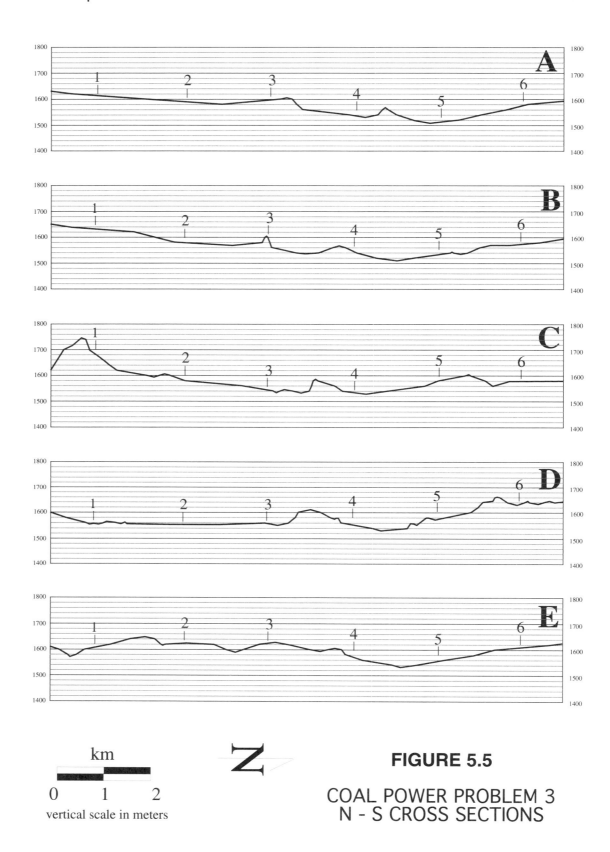

FIGURE 5.5

COAL POWER PROBLEM 3
N - S CROSS SECTIONS

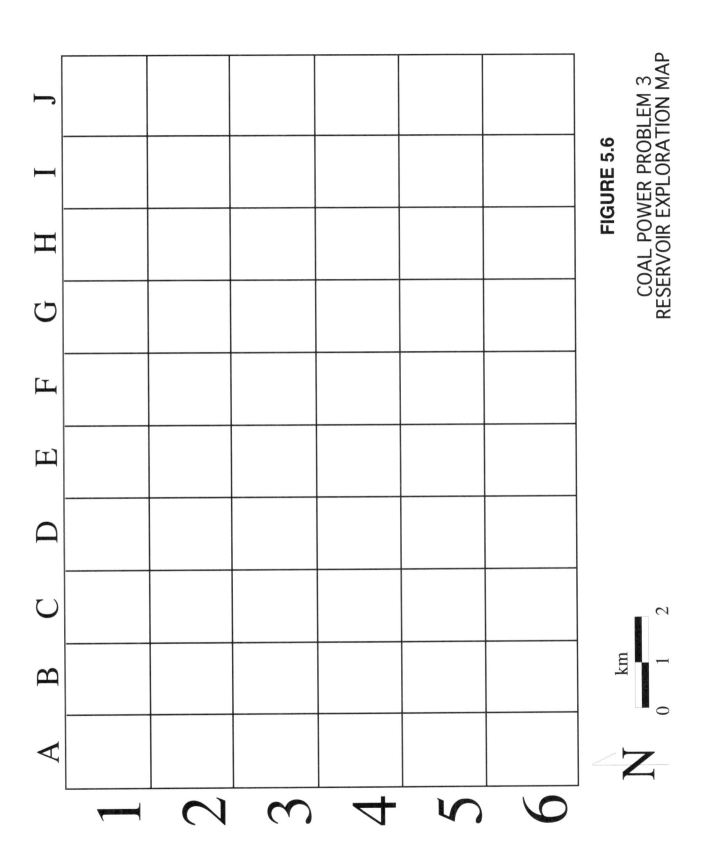

FIGURE 5.6
COAL POWER PROBLEM 3
RESERVOIR EXPLORATION MAP

186 Chapter 5: Mire to Fire

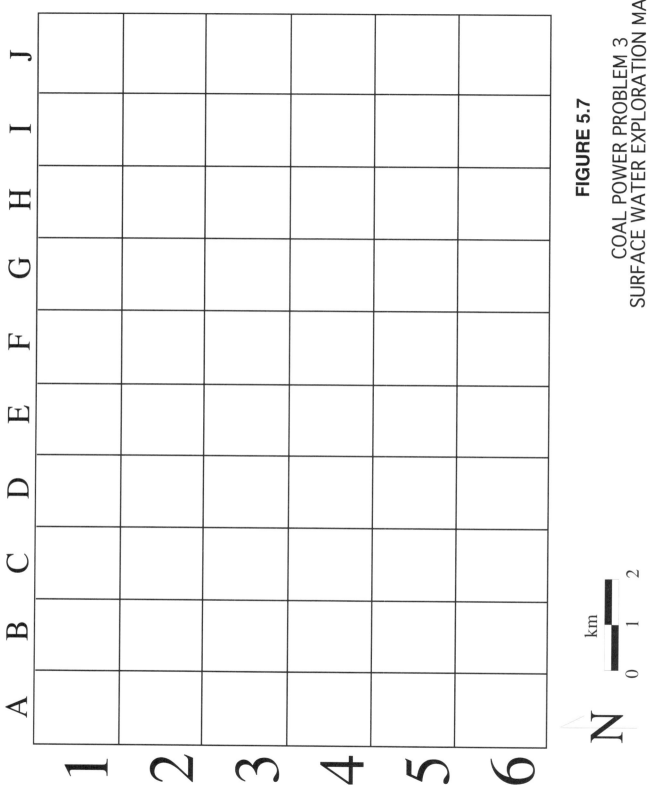

FIGURE 5.7
COAL POWER PROBLEM 3
SURFACE WATER EXPLORATION MAP

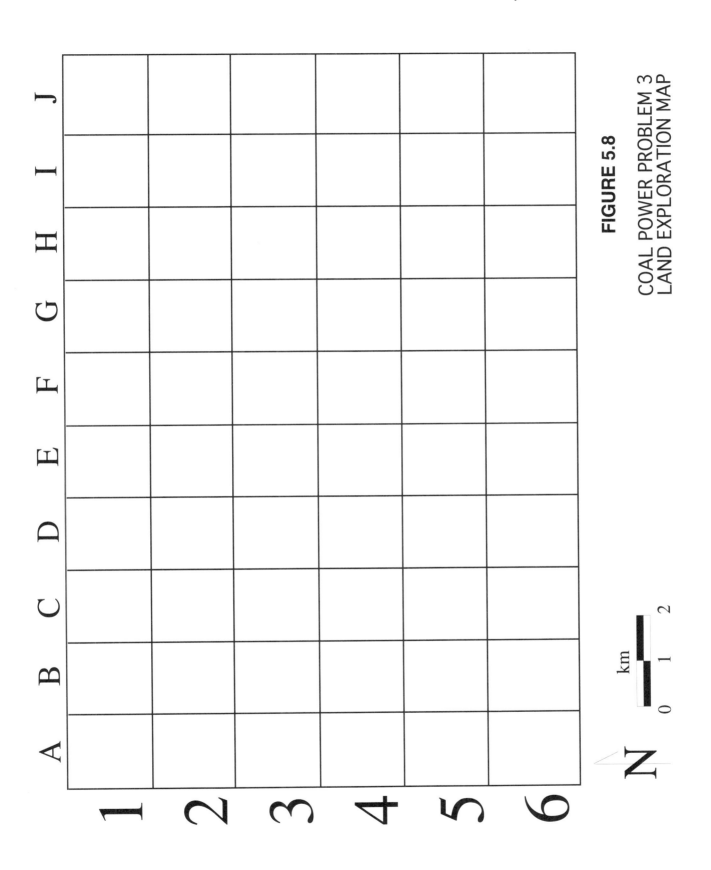

FIGURE 5.8
COAL POWER PROBLEM 3
LAND EXPLORATION MAP

Coal Power Research Problem 4: Operating a Coal-Fired Power Plant

Introduction and Advice

Before you start the research, examine **Study Notes 5.10** and **5.11** for some guidance about what you need to know before you start and what you should know when you are done. In **Study Note 5.12**, we provide a table you can use to keep track of assignments and their due dates.

> **Study Note 5.10: As you begin the research, you will need to have some prior knowledge in the following areas:**
>
> 1. An understanding of coal formation and coal quality.
> 2. Background on use of coal to generate electricity.
> 3. Familiarity with coal combustion chemistry.

> **Study Note 5.11: What you should be able to do after completing the research.**
>
> 1. Learn how a power plant uses coal to make electricity.
> 2. Understand the basics about coal-fired power plant systems, their limitations, and use.
> 3. Be knowledgeable about factors shaping efficient operations of a power plant.
> 4. Become comfortable with complex systems by adjusting power plant systems to maintain electrical output when stresses are applied.
> 5. Understand the chemistry of the scrubber reactions.

Study Note 5.12: Use the table provided below to list the research assignments given by your instructor. Make notes on this table about the types of work you will be expected to submit for a grade.

Research Assignments	Specific Instructions for Completing Assignment	Completion Date
Phase 1		
Phase 2		
Phase 3		
Other Assignments		

What's the Problem?

What would the world be like without using coal for generation of electricity and heat? You can check a variety of references and probably conclude that coal-fired power plants produce about 25% of human energy needs. There are both pros and cons to the use of coal, and the general topic of coal's use for power generation has caused heated debate.

For example, there are issues related to disturbance of habitats from mining activities, health problems related to the mining itself, transportation of coal, storage, cleaning, siting of coal-fired power plants, and political issues related to where coal is found and the economic advantage it provides to countries with large coal reserves. With the increasing world population, society is demanding the production of more energy. Coal is relatively abundant on the Earth, and the estimated world reserves could power the world for centuries into the future. However, is coal a good choice for meeting increased energy demands? How would you make a decision about financing use of coal instead of other sources of energy?

Coal power has been in use for long enough that many of the environmental impacts have been studied. We have also reached a point at which mitigation measures have been designed and implemented. Even so, combustion of coal is responsible for half of the sulfur dioxide and about a third of the nitrogen oxides that humans produce. These gases play important roles in the atmospheric chemistries that lead to acid rain.

In this virtual world, we will give you a chance to run a coal-fired power plant. This is an important exercise because it gives you a chance to study the complexity of the power plant operation, particularly the many variables that must be adjusted simultaneously in order to increase, decrease, or maintain levels of electricity generation. We do not take a position about whether coal is a good or bad choice of fuel for generating electricity. Rather, we give you a complicated problem and ask you to work through the details. Later, you can look at your experience and be able to think more deeply about the advantages and disadvantages of using coal in electrical power plants.

Background Information

To begin the operation of a coal-fired power plant you will need to become familiar with the systems that are generally found in such plants. There are 5 main systems:
- The Cooling Water System
- The Coal Storage and Delivery System
- The Furnace System
- The Baghouse System
- The Scrubber and Stack System

We will spend some time examining each of these systems so you understand them well enough to later adjust them while operating the power plant.

The Cooling Water System (CPS)

The Cooling Water System comprises a pump array and a large, elongate reservoir that holds 6 billion gallons of water. Water from the reservoir is pumped from the inlet end of the lake into the Furnace System. The reservoir water cools the furnace steam (a completely separate fluid system) so it may be cycled through the furnace again. The reservoir water leaves the Furnace System about 25 degrees hotter than when it entered, and it is routed to the opposite end of the reservoir. As it slowly moves towards the inlet again, it looses heat to the atmosphere and cools down enough to be used again. The water temperature at the inlet depends on the ambient air temperature. In the summer it can be as hot as 90° Fahrenheit, while during the winter it can be as cool as 50° F.

The Coal Storage and Delivery System (CSDS)

Approximately 7 days of fuel is maintained at the power plant in the Coal Storage and Delivery System. Three times a day, coal arrives from the nearby mine via electric rail, where it is stored in a storage yard. Increased demands require several weeks before a new coal production level from the mine is achieved.

The coal is stored in long piles, and a movable retrieval system is introduced into these piles as needed. The retrieval system is basically a conveyor belt that moves coal to the Furnace System.

The Furnace System (FS)

In many ways, the Furnace System is both the most primitive and the most complicated part of the plant. The **furnace** is the chamber where coal is burned. The coal is prepared for burning by pulverizing it under gigantic revolving steel wheels. Fans combine air with pulverized coal, pressurize the mixture, and force it into the furnace chamber through burners or nozzles. Uniform heating of the furnace chamber is achieved by aiming the **burners** tangential to the chamber, creating a "cyclonic" fireball. Other fans are set to force the fly ash and combustion gases out of the chamber towards the Baghouse and Scrubber/Stack Systems.

Operators have to be careful not to cause a **flame out**, which occurs when the furnace fireball is extinguished from too rich a fuel mixture (not enough O_2), or through some other flame extinguishing event such as rupturing of steam pipes in the furnace. Following a flame out, there is a chance for a furnace explosion. Basically, oxygen levels creep up to the levels required by the furnace and unburned fuel ignites in an uncontrolled fashion. As you might guess, an explosion can devastate a power plant.

The furnace is the only place in the power plant where the production of the acid rain-producing gases, nitrogen oxides (NO and NO_2), can be controlled. Nitrogen oxides are the products of an oxygen-rich setting during combustion in the furnace. You can minimize nitrogen oxides by carefully adjusting the amount of oxygen mixed with the coal at the burners.

Specific furnace components include:
- **Boiler** — a network of pipes and chambers along the inner walls of the furnace chamber. The boiler contains very pure water that is first heated to steam and then superheated (temperature increased while not in equilibrium with water). This steam is then directed through turbines.

- **Turbine** — a windmill-like device with hundreds of blades that spin in response to a pressure differential provided by the superheated steam entering one end, doing work, and then exiting at a lower pressure. The turbine, in turn, spins a generator (always at 3,600 rpm) which produces the electrical power. By adjusting the direct current load in the generator (always maintaining 3600 rpm) more power can be generated. The generator must spin at 3600 rpm to provide electricity at the correct frequency for the regional power grid (60 cycles or 60 Hz)
- **Condenser** — a device that removes heat from the steam, reverting it to water. This is done by running cool water from the Cooling Water System reservoir around a network of pipes that contain the low pressure steam as it leaves the turbine. As the heat from the steam is transferred to the cooling water, the steam condenses into water. The water takes up less volume, a side effect that occurs as the steam condenses, which creates a low pressure zone that, in turn, aids the operation of the turbine.

The Baghouse System (BC)

The **baghouse** is the location in the plant that captures the solid residues of combustion. Coal has a non-burnable fraction called ash. When coal is consumed in the furnace chamber, all that remains is a gray-black residue of sand-size particles. This ash is considered a pollutant both because of its particulate nature (lung irritant) and the fact it contains heavy metals. The power plant needs to remove 99.9 % of the ash from the air exiting the plant.

The baghouse is a series of 24 rooms, each of which contain about 800 fabric tubes or bags. These hang vertically in the **baghouse building**. The pressurized gases and fly ash from the furnace are blown into the bags, allowing the gases to escape while the fly ash is captured. Individual baghouse rooms can be isolated and the bags emptied. The fly ash is moved from the baghouse room to a transfer station at the plant. From there the fly ash is taken via trucks back to the mine for its ultimate disposal (back into the hole that the coal was extracted from).

The Scrubber and Stack System (SSS)

The Scrubber and Stack System removes some of the acid rain producing gas sulfur dioxide (SO_2). Some of the gases resulting from the combustion of coal include sulfur dioxide, nitrogen oxides (NO_x), carbon dioxide, and carbon monoxide.

The power plant is required by the EPA to remove 73% of the sulfur dioxide (SO_2) from the furnace gases. The removal is accomplished at our power plant through a process called **scrubbing**. Scrubbing involves moving about 75% of the furnace gases through enormous tanks where a mist of calcium hydroxide solution is spraying. The calcium hydroxide reacts with the SO_2 gas and forms a solid (calcium sulfite). The calcium sulfite falls out of the air and accumulates on the bottom of the tank. The cleaned gas is mixed back in with unscrubbed gas and exits through the stack. Sensors on the stack relay the amount of SO_2 and NO_x that are leaving the plant.

Research Questions

We are not going to pose any framework of research questions for this simulation. We encourage you to work with classmates and your faculty to decide if it is important to establish specific questions, and then conduct research activities that will yield the data necessary to answer these questions. You can certainly use the question-driven approach, as we did in most of the other research simulations.

On the other hand, we sometimes don't know enough to ask meaningful questions about a system. We may have some tentative ideas about how a system works and, therefore, we observe the system to see if we can answer some of our questions, or discover features of the system that allow us to refine our questioning and, sometimes, set up descriptive or experimental research projects. For this simulation, we encourage you to use a systematic approach to study the individual systems of the power plant. Then you can try your hand at operating the power plant under normal conditions and conditions that stress the system.

Phases of Research

To provide at least some guidance, we suggest you explore the power plant by pursuing three phases of work. First, survey the various systems in the plant, and answer the questions we provide for each system. Second, try integrating the systems and operate the whole plant. Third, and finally, play with the simulation of stressed conditions when the power plant must suddenly increase its output of electricity.

Phase 1 — Part 1: Survey of the systems.

Remember that to begin the operation of power plant, you will need to become familiar with each system separately. The 5 main systems are:
- The Cooling Water System
- The Coal Storage and Delivery System
- The Furnace System
- The Baghouse System
- The Scrubber and Stack System

Click on the image of the power plant to enter the simulation. Then, run the arrow over the different parts of the scene to locate the five systems we have described. Clicking on any system selects it and displays its operation.

Task 1:
Begin by activating the Cooling Water System (CPS). The pumping rate of the cooling water system can be adjusted to accommodate the needs of the Furnace System. Try adjusting the pumping rate now and observe how the system behaves. Answer the following questions.

a) What happens to the temperature of the reservoir water at the inlet end as you adjust the pumping rate, and why?

_____

_____

b) Considering the seasonal difference of water temperature at the inlet is pumping at different rates significant? Explain why or why not ?

Task 2:

Now, activate the Coal Storage and Delivery System (CSDS). The delivery of coal from the mine can be adjusted, as well as the delivery rate of pulverized coal to furnace. Go ahead and adjust these parameters now. Then, answer the following questions:

a) What are the consequences of increasing the delivery rate to the furnace without changing the mine delivery rate?

b) What happens if you increase the delivery rate from the mine without increasing the furnace delivery rate?

Task 3:

Next, activate the Furnace System (FS) works. Adjust the various settings to explore how the Furnace System works. You may want to review the different furnace components.

Task 4:

Activate the Baghouse System (BC). Remember, this is the system that captures the solid residues of combustion. The rate at which rooms in the baghouse are emptied can be controlled by varying the number of rooms on line. Try adjusting this rate now, and then answer the following question.

What happens to the overall fly-ash removal amount as you take rooms off line for repair?

Task 5:

Finally, we turn to the Scrubber and Stack System (SSS). You can adjust the amount of furnace gas that runs through the scrubber. Try adjusting this amount. From your observations of what happens as you vary the controls, answer the following questions.

a) What happens to the values for SO_2 exiting the stack when you adjust the diverters?

b) What happens to the amount of calcium hydroxide used as you adjust the diverters, and why?

c) Why doesn't the power plant divert all the gas through the scrubbers?

Phase 2 — Operating the whole plant.

Now that you are familiar with the sub-systems you can try operating the entire plant. Before you start, review the way each system works. From this review, develop a strategy for operating all of the systems together. For example, how do you start up a coal plant? You obviously have to light the furnace before you can generate steam to drive the turbines to make power.

Task 1:
Once again, think about the different systems. What systems would have to be operational before you can light the furnace?

_____
_____
_____
_____

Task 2:
Based on what you discovered completing *Task 1*, you will have to turn on the other systems before you start the furnace. Do that now. This power plant has a start up feature that uses natural gas to pre-heat the furnace, so when you introduce the coal, it will burn properly. If you're convinced that you have turned on the appropriate sub-systems, then go ahead and begin the start up procedure.

Task 3:
Once the correct temperatures are achieved (the system will tell you), you can begin to add fuel until you have reached operating levels. The normal operating level is an electricity production rate of 720,000 kilowatts.

Task 4:
The Baghouse and Scrubber Systems start up at certain levels that may not be appropriate to running the power plant at maximum efficiency. Make sure you adjust these systems so you are reaching your goals.

Task 5:
For the Scrubber System, the goal is to remove 74% of the SO_2 found in the unscrubbed furnace gas. We will have to try to remove 99.9% of the ash. That is exactly what happens if the furnace gases pass through the baghouse. There is no way to pump fly-ash laden furnace gases directly out the stack because the scrubber system is between the stack and the baghouse. If the baghouse becomes full, then the entire system will shut down to save the scrubber facility. Don't let this happen! From your observations, answer the following questions:

a) Why do you think it would be a catastrophe if fly-ash was pumped into the scrubbers?

_____
_____

Chapter 5: Mire to Fire

b) Now you should adjust the pollution control systems so you are reaching your regulatory levels. Describe your nitrogen oxide levels, and provide a rationale for reducing them by operating the burners in the furnace to use less oxygen.

_____
_____
_____
_____

Phase 3 — Need a little stress in your life?

A nuclear power plant in California has gone off-line after an earthquake. The authorities have asked if you can bump up your power production to 810,000 kilowatts to help meet the needs of Phoenix and Los Angeles.

In the Furnace System, find the DC load controls and ramp up the load to the requested amount. You'll notice the generators can fall below 3,600 rpm and the power grid goes off-line because of the incompatibility of the power plant alternating current with the regional grid frequency.

Task 1:
Try to bring the power plant back on-line by adjusting the other systems, while staying in your pollution control limits. Respond to the following questions:

a) Explain what you do to increase the rpm of the turbines.

_____
_____
_____
_____

b) Decide which sub-systems you have to adjust and describe your rationale.

_____
_____
_____
_____

In the Know about Operating a Power Plant

After running the simulation, you believe you know a fair amount about the many variables shaping the operations of a coal-fired power plant. Unfortunately, you discuss the simulation with an old friend who worked as an engineer building a power plant with similar systems. She tells you that simulations of plant operations are very difficult to design, because you need a pretty sophisticated modeling program to mimic the complexities of the many interacting and often interdependent variables. She points out you have been using a simulation in which you don't even know the equations that were used to calculate electrical output from a combination of variables.

Your friend's criticism worries you, especially since you believed you had gained some important insight into power plant operations. In particular, you felt you could at least make intelligent decisions on environmental issues related to power plant operations. You make a list of your friend's major points and decide to do some research on each area so you can then decide what you really know and don't know after working through the simulation. Here is the list of criticisms from your friend.

1. Start with some basics. The coal seam appears to be of a quality and quantity that will support the plant's operation for 20 years. However, since no one could look at all of the coal, what are the data that suggest you have the amount and type of coal you need? What is the level of confidence you have in the conclusions you have drawn from the data?

2. Okay, so you are pretty sure the coal source is adequate. Remember that each power plant has multiple systems. Each system has multiple variables that shape the outcome of the system's functioning. How can you develop a model for each of the single systems and then combine the models of multiple systems to get a composite model that mimics how a power plant works?

3. Even if you can model all of the systems and their interactions as a composite, how does your model changes through time? For example, how can you predict power usage from the population the power plant serves? Suppose the population stays the same, but their lifestyles lead to increases in use of electrical appliances. Or, suppose that the regional population increases, and their use of electricity also increases. How can you accurately predict such changes through time and incorporate these predictions into your power plant simulation?

4. Why even bother with simulations, given the difficulty in designing and calibrating them?

SO, WHAT'S THE POINT?

You are a professor in a college located just outside the town in the area where the utility company is trying to locate a suitable site for a coal-burning power plant. Since your training is in environmental geology, and you are an expert in the regional geological features, the utility company invites you to work as a consultant with them. They assign you to work with the team that is locating both the coal seam and potential cooling water reservoirs. They also ask you to be one of a four-person review panel who will examine surface water rights and patterns of land use in the region. In short, the utility company feels it will benefit from your knowledge, and also from your input as a resident of the area.

You agree, but immediately many of your colleagues and your neighbors in town express their concerns about you contributing to the ruin of the countryside with a coal-fired power plant. In your opinion, you did not believe there was much anyone in the area could do about stopping the building of the power plant. In fact, you felt your participation in the project would ensure a local resident would have some substantive input into various facets of the decision processes for siting the plant and its water reservoirs, as well as on the determinations of surface water issues and land use patterns.

Quite a quandry! You feel your role is important to the area, but colleagues and neighbors are really upset with you. After thinking deeply about what you know and don't know, you finally outline a plan for gathering information. This plan is a list of questions you feel you need to answer for yourself. Answer these questions, and decide if you made the correct decision in accepting the role you think of as both consultant and protector of the area in which you live.

1. Does a local population have any power to stop the building of a power plant scheduled for siting within their area? If we do have such power, when should we exercise it?
2. What kind of power plant has been planned for the area? Of course, it will use coal, but how efficient is it, how long has this type of model been in operation, have there been problems with this model in the past, and how does it compare to other models used elsewhere in the United States and in other countries?
3. What are the environmental impacts in areas where it has been used? How do I think about environmental issues, broadly defined to include special habitats, endangered species, vegetation in the region, animals, water systems, landforms, and aesthetically pleasing vistas.
4. Are there likely to be impacts on humans? For example, are there health risks to humans from the mining of the coal and the operation of the plant?
5. Will my family and neighbors benefit from the new jobs available at the plant and, perhaps, improvements in our economy?
6. Will taxes on a new industry provide the town and county with funds for road improvements, schools, and support of emergency systems (police, fire departments, and medical services)?

Table 5.3
Ash Composition

Locality	Clay	Quartz	Hematite	Carbonates	Sulfates	Pyrite
row col						
row col						
row col						
row col						
row col						
row col						
row col						
row col						
row col						
row col						
row col						
row col						
row col						
row col						
row col						
row col						
row col						
row col						
row col						
row col						

Table 5.3
Ash Composition

Locality	Clay	Quartz	Hematite	Carbonates	Sulfates	Pyrite
row col						
row col						
row col						
row col						
row col						
row col						
row col						
row col						
row col						
row col						
row col						
row col						
row col						
row col						
row col						
row col						
row col						
row col						
row col						
row col						